INTERNATIONAL STANDARD

ISO/IEC 18004

Third edition
2015-02-01

Information technology — Automatic identification and data capture techniques — QR Code bar code symbology specification

Technologies de l'information — Technologie d'identification automatique et de capture des données — Spécification de la symbologie de code à barres Code QR

Reference number
ISO/IEC 18004:2015(E)

Contents

Page

Foreword

ISO (the International Organization for Standardization) and IEC (the International Electrotechnical Commission) form the specialized system for worldwide standardization. National bodies that are members of ISO or IEC participate in the development of International Standards through technical committees established by the respective organization to deal with particular fields of technical activity. ISO and IEC technical committees collaborate in fields of mutual interest. Other international organizations, governmental and non-governmental, in liaison with ISO and IEC, also take part in the work. In the field of information technology, ISO and IEC have established a joint technical committee, ISO/IEC JTC 1.

The procedures used to develop this document and those intended for its further maintenance are described in the ISO/IEC Directives, Part 1. In particular the different approval criteria needed for the different types of document should be noted. This document was drafted in accordance with the editorial rules of the ISO/IEC Directives, Part 2 (see www.iso.org/directives).

Attention is drawn to the possibility that some of the elements of this document may be the subject of patent rights. ISO and IEC shall not be held responsible for identifying any or all such patent rights. Details of any patent rights identified during the development of the document will be in the Introduction and/or on the ISO list of patent declarations received (see www.iso.org/patents).

Any trade name used in this document is information given for the convenience of users and does not constitute an endorsement.

For an explanation on the meaning of ISO specific terms and expressions related to conformity assessment, as well as information about ISO's adherence to the WTO principles in the Technical Barriers to Trade (TBT) see the following URL: Foreword - Supplementary information

The committee responsible for this document is ISO/IEC JTC 1, *Information technology*, SC 31, *Automatic identification and data capture techniques*.

This third edition cancels and replaces the second edition (ISO/IEC 18004:2006), which has been technically revised.

Introduction

It is necessary to distinguish four technically different, but closely related members of the QR Code family, which represent an evolutionary sequence.

- QR Code Model 1 was the original specification for QR Code and is described in AIM ITS 97-001 International Symbology Specification-QR Code.
- QR Code Model 2 was an enhanced form of the symbology with additional features (primarily the addition of alignment patterns to assist navigation in larger symbols), and was the basis of the first edition of ISO/IEC 18004.
- QR Code (the basis of the second edition of ISO/IEC 18004) is closely similar to QR Code Model 2, its QR Code format differs only in the addition of the facility for symbols to appear in a mirror image orientation for reflectance reversal (light symbols on dark backgrounds) and the option for specifying alternative character sets to the default.
- The Micro QR Code format (also specified in the second edition of ISO/IEC 18004), is a variant of QR Code with a reduced number of overhead modules and a restricted range of sizes, which enables small to moderate amount of data to be represented in a small symbol, particularly suited to direct marking on parts and components, and to applications where the space available for the symbol is severely restricted.

QR Code is a matrix symbology. The symbols consist of an array of nominally square modules arranged in an overall square pattern, including a unique finder pattern located at three corners of the symbol (in Micro QR Code symbols, at a single corner) and intended to assist in easy location of its position, size, and inclination. A wide range of sizes of symbol is provided for, together with four levels of error correction. Module dimensions are user-specified to enable symbol production by a wide variety of techniques.

QR Code Model 2 symbols are fully compatible with QR Code reading systems.

Model 1 QR Code symbols are recommended only to be used in closed system applications and it is not a requirement that equipment complying with this International Standard should support Model 1. Since QR Code is the recommended model for new, open systems application of QR Code, this International Standard describes QR Code fully, and lists the features in which Model 1 QR Code differs from QR Code in Annex N.

4 Terms and definitions

For the purposes of this document, the terms and definitions given in ISO/IEC 19762-1 and ISO/IEC 19762-2 and the following apply.

4.1
character count indicator
bit sequence which defines the data string length in a mode

4.2
data masking
process of XORing the bit pattern in the encoding region with a data mask pattern to provide a symbol with more evenly balanced numbers of dark and light modules, and reduced occurrence of patterns which would interfere with fast processing of the image

4.3
data mask pattern reference
three-bit identifier of the data masking patterns applied to the symbol

4.4
encoding region
region of the symbol not occupied by function patterns and available for encoding of data and error correction codewords, and for Version and format information

4.5
exclusive subset
subset of characters within the character set of a mode which are not shared with the more restricted character set of another mode

4.6
extension pattern
function pattern in Model 1 symbols, which does not encode data

4.7
format information
encoded pattern containing information on symbol characteristics essential to enable the remainder of the encoding region to be decoded

4.8
QR Code
pertaining to QR Code symbols identified as versions 1 to 40, as distinct from Micro QR Code symbols

4.9
function pattern
overhead component of the symbol (finder, separator, timing patterns, and alignment patterns) required for location of the symbol or identification of its characteristics to assist in decoding

4.10
masking
process of XORing the bit pattern in an area of the symbol with a mask pattern to reduce the occurrence of patterns which would interfere with fast processing of the image

4.11
micro
pertaining to Micro QR Code symbols identified as versions M1 to M4, as distinct from QR Code symbols

4.12
mode
method of representing a defined character set as a bit string

4.13
mode indicator
four-bit identifier indicating in which mode the following data sequence is encoded

4.14
padding bit
zero bit, not representing data, used to fill empty positions of the final codeword after the terminator in a data bit string

4.15
remainder bit
zero bit, not representing data, used to fill empty positions of the symbol encoding region after the final symbol character, where the area of the encoding region available for symbol characters does not divide exactly into 8-bit symbol characters

4.16
remainder codeword
pad codeword, placed after the error correction codewords, used to fill empty codeword positions to complete the symbol if the total number of data and error correction codewords does not exactly fill its nominal capacity

4.17
segment
sequence of data encoded according to the rules of one ECI or encoding mode

4.18
separator
function pattern of all light modules, one module wide, separating the finder patterns from the rest of the symbol

4.19
symbol number
three-bit field indicating the symbol version and error correction level applied, used as part of the format information in Micro QR Code symbols

4.20
terminator
bit pattern of defined number (depending on symbol) of all zero bits used to end the bit string representing data

4.21
timing pattern
alternating sequence of dark and light modules enabling module coordinates in the symbol to be determined

4.22
version
size of the symbol represented in terms of its position in the sequence of permissible sizes for Micro QR Code symbols from 11 × 11 modules (version M1) to 17 × 17 modules (version M4) or, for QR Code symbols, from 21 × 21 modules (version 1) to 177 × 177 (version 40) modules

Note 1 to entry: The error correction level applied to the symbol may be suffixed to the version designation, e.g. version 4-L or version M3-Q.

4.23
version information
encoded pattern in certain QR Code symbols containing information on the symbol version together with error correction bits for this data

5 Mathematical and logical symbols, abbreviations and conventions

5.1 Mathematical and logical symbols

Mathematical symbols used in formulae and equations are defined after the formula or equation in which they appear.

For the purposes of this document, the following mathematical operations apply.

div is the integer division operator;

mod is the integer remainder after division;

XOR is the exclusive-or logic function whose output is one only when its two inputs are not equivalent. It is represented by the symbol $\oplus$.

5.2 Abbreviations

BCH Bose-Chaudhuri-Hocquenghem

ECI Extended Channel Interpretation

RS Reed-Solomon

5.3 Conventions

5.3.1 Module positions

For ease of reference, module positions are defined by their row and column coordinates in the symbol, in the form (i, j) where i designates the row (counting from the top downwards) and j the column (counting from left to right) in which the module is located, with counting commencing at 0. Module (0, 0) is therefore located at the upper left corner of the symbol.

5.3.2 Byte notation

Byte contents are shown as hex values.

5.3.3 Version references

For QR Code symbols, symbol versions are referred to in the form Version V-E where V identifies the version number (1 to 40) and E indicates the error correction level (L, M, Q, H).

For Micro QR Code symbols, symbol versions are referred to in the form Version MV-E where the letter M indicates the Micro QR Code format and V (with a range of 1 to 4) and E (with values L, M and Q) have the meanings defined above.

6 Symbol description

6.1 Basic characteristics

QR Code is a matrix symbology with the following characteristics:

a) Formats:

1) QR Code, with full range of capabilities and maximum data capacity;

   2) Micro QR Code, with reduced overhead, some restrictions on capabilities and reduced data capacity (compared with QR Code symbols).

b) Encodable character set:

   1) numeric data (digits 0 - 9);

   2) alphanumeric data (digits 0 - 9; upper case letters A - Z; nine other characters: space, $ % * + - . / :);

   3) byte data [default: ISO/IEC 8859-1; or other sets as otherwise defined (see 7.3.5)];

   4) Kanji characters. Kanji characters in QR Code can be compacted into 13 bits.

c) Representation of data:

   A dark module is nominally a binary one and a light module is nominally a binary zero. However, see 6.2 for details of reflectance reversal.

d) Symbol size (not including quiet zone):

   1) Micro QR Code symbols: 11 × 11 modules to 17 × 17 modules (Versions M1 to M4, increasing in steps of two modules per side);

   2) QR Code symbols: 21 × 21 modules to 177 × 177 modules (Versions 1 to 40, increasing in steps of four modules per side).

e) Data characters per symbol

   1) maximum Micro QR Code symbol size, Version M4-L):

      — numeric data: 35 characters

      — alphanumeric data: 21 characters

      — Byte data: 15 characters

      — Kanji data: 9 characters

   2) maximum QR Code symbol size, Version 40-L:

      — numeric data: 7 089 characters

      — alphanumeric data: 4 296 characters

      — Byte data: 2 953 characters

      — Kanji data: 1 817 characters

f) Selectable error correction:

   Four levels of Reed-Solomon error correction (referred to as L, M, Q and H in increasing order of capacity) allowing recovery of:

   — L 7%

   — M 15%

   — Q 25%

   — H 30%

   of the symbol codewords.

For Micro QR Code symbols, error correction level H is not available. For Version M1 Micro QR Code symbols, the RS capacity is limited to error detection only.

g) Code type:

Matrix

h) Orientation independence:

Yes (both rotation and reflection)

Figure 1 illustrates a Version 1 QR Code symbol in normal colour and with reflectance reversal (see 6.2), in both normal and mirror image orientations.

Figure 2 illustrates a Version M2 Micro QR Code symbol in normal colour and with reflectance reversal (see 6.2), in both normal and mirror image orientations.

6.2 Summary of additional features

The use of the following additional features is optional in QR Code:

— Structured append

This allows files of data to be represented logically and continuously in up to 16 QR Code symbols. These may be scanned in any sequence to enable the original data to be correctly reconstructed. Structured Append is not available with Micro QR Code symbols.

— Extended Channel Interpretations

This mechanism enables data using character sets other than the default encodable set (e.g. Arabic, Cyrillic, Greek) and other data interpretations (e.g. compacted data using defined compression schemes) or other industry-specific requirements to be encoded. Extended Channel Interpretations other than the default interpretation are not available in Micro QR Code symbols.

— Reflectance reversal

Symbols are intended to be read when marked so that the image is either dark on light or light on dark (see Figures 1 and 2). The specifications in this International Standard are based on dark images on a light background, therefore in the case of symbols produced with reflectance reversal references to dark or light modules should be taken as references to light or dark modules respectively.

— Mirror imaging

The arrangement of modules defined in this International Standard represents the "normal" orientation of the symbol. It is, however, possible to achieve a valid decode of a symbol in which the arrangement of the modules has been laterally transposed. When viewed with the finder patterns at the top left, top right and bottom left corners of the symbol, the effect of mirror imaging is to interchange the row and column positions of the modules.

(a) (b)

(c) (d)

Figure 1 — Examples of QR Code symbol encoding the text "QR Code Symbol" - (a) normal orientation and normal reflectance arrangement; (b) normal orientation and reversed reflectances; (c) mirror image orientation and normal reflectance arrangement; (d) mirror image orientation and reversed reflectances

(a) (b)

(c) (d)

Figure 2 — Examples of Version M2 Micro QR Code symbol encoding the text "01234567" - (a) normal orientation and normal reflectance arrangement; (b) normal orientation and reversed reflectances; (c) mirror image orientation and normal reflectance arrangement; (d) mirror image orientation and reversed reflectances

NOTE The corner marks in Figures 1 and 2 indicate the extent of the quiet zone.

6.3 Symbol structure

6.3.1 General

Each QR Code symbol shall be constructed of nominally square modules set out in a regular square array and shall consist of an encoding region and function patterns, namely finder, separator, timing patterns, and alignment patterns. Function patterns do not encode data. The symbol shall be surrounded on all four sides by a quiet zone border. Figure 3 illustrates the structure of a Version 7 symbol. Figure 4 illustrates the structure of a Version M3 symbol.

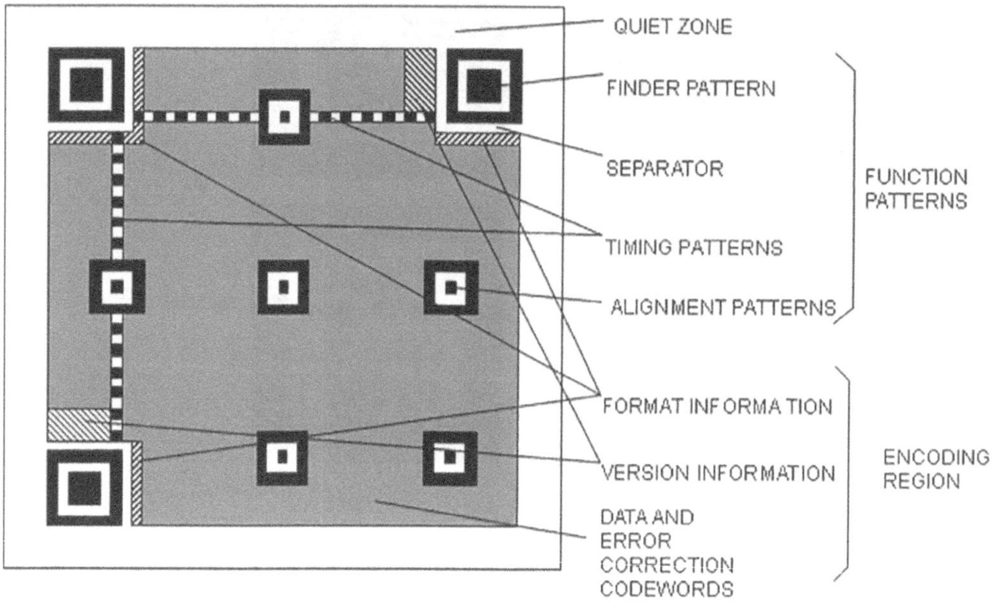

Figure 3 — Structure of a QR Code symbol

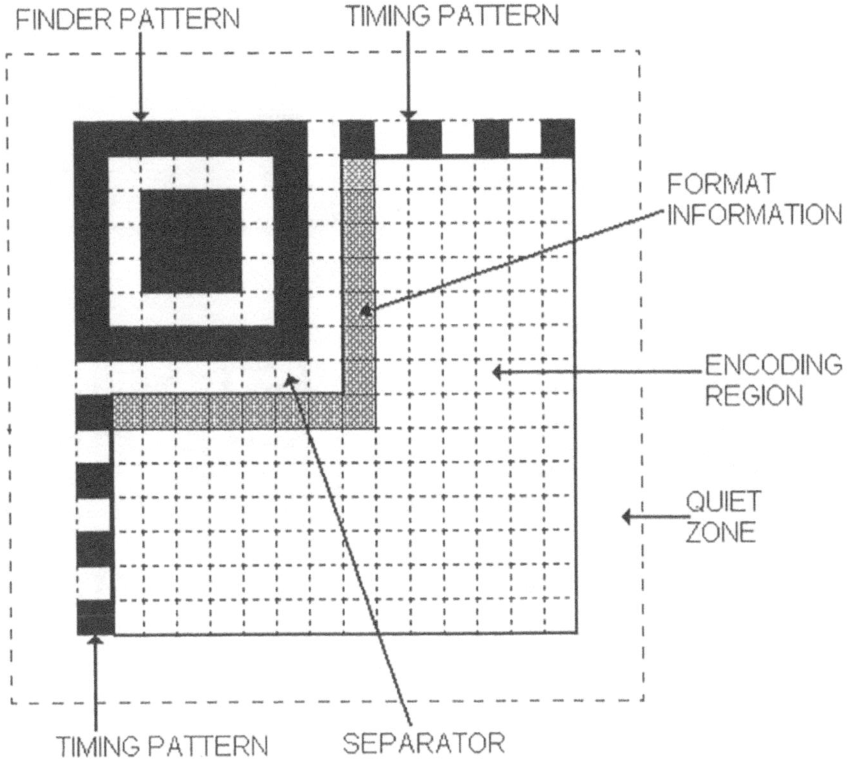

Figure 4 — Structure of Version M3 Micro QR Code symbol

6.3.2 Symbol Versions and sizes

6.3.2.1 QR Code symbols

There are forty sizes of QR Code symbol referred to as Version 1, Version 2 ... Version 40. Version 1 measures 21 modules × 21 modules, Version 2 measures 25 modules × 25 modules and so on increasing in steps of 4 modules per side up to Version 40 which measures 177 modules × 177 modules. Figures 5 to 10 illustrate the structure of Versions 1, 2, 6, 7, 14, 21 and 40.

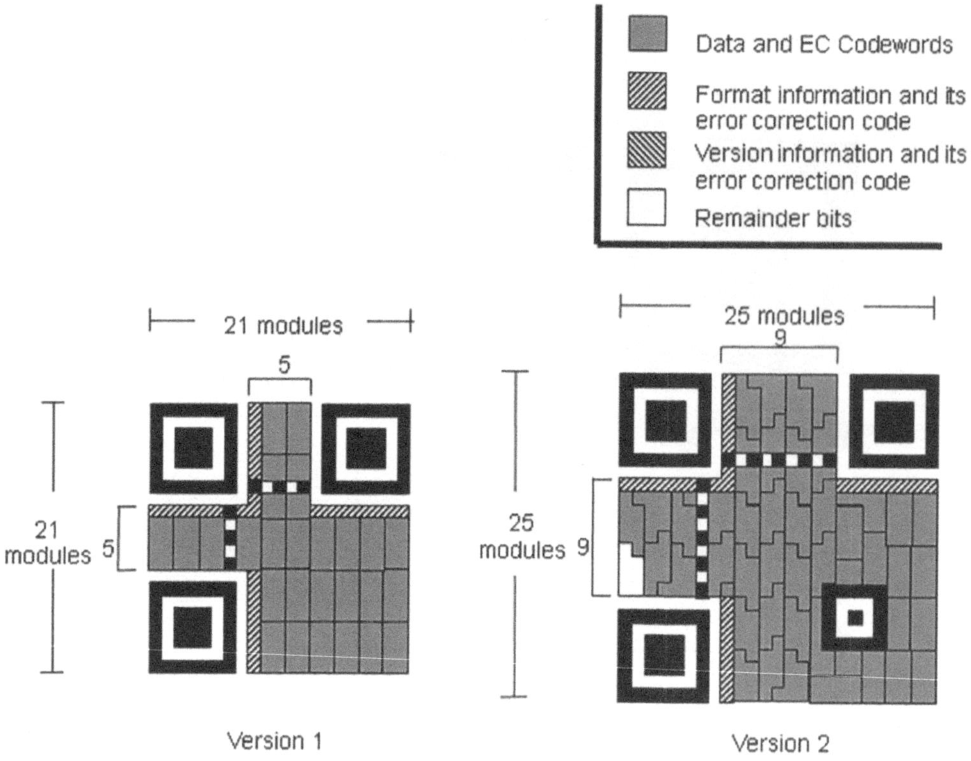

Figure 5 — Version 1 and 2 symbols

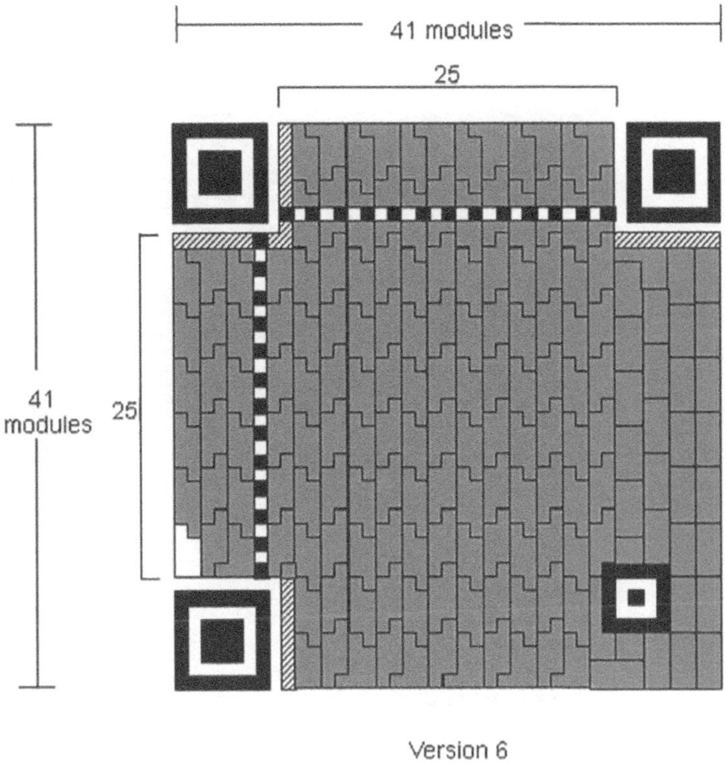

Figure 6 — Version 6 symbol

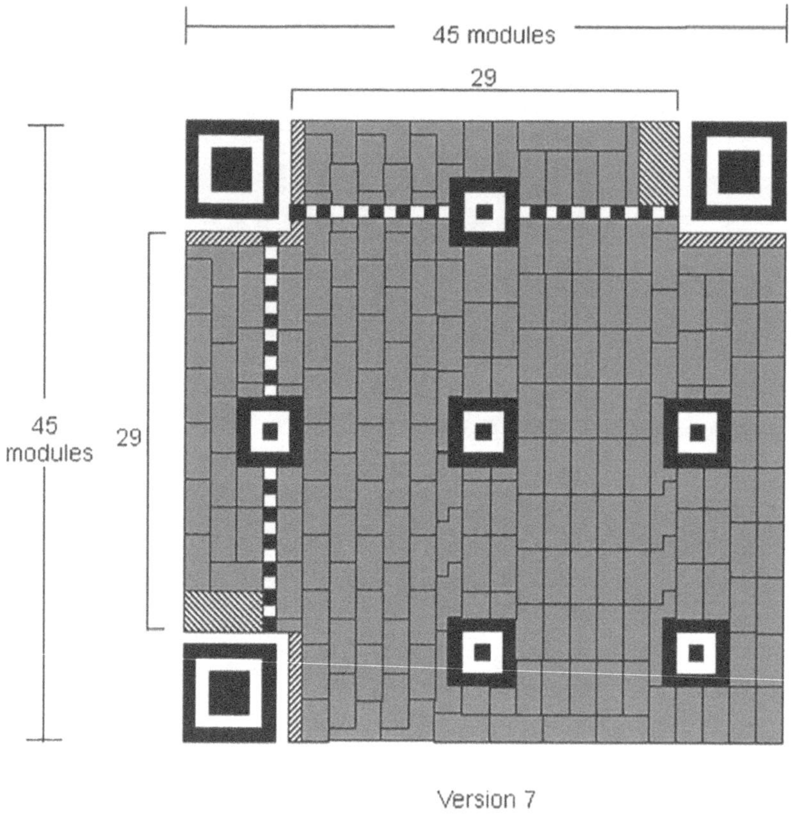

Figure 7 — Version 7 symbol

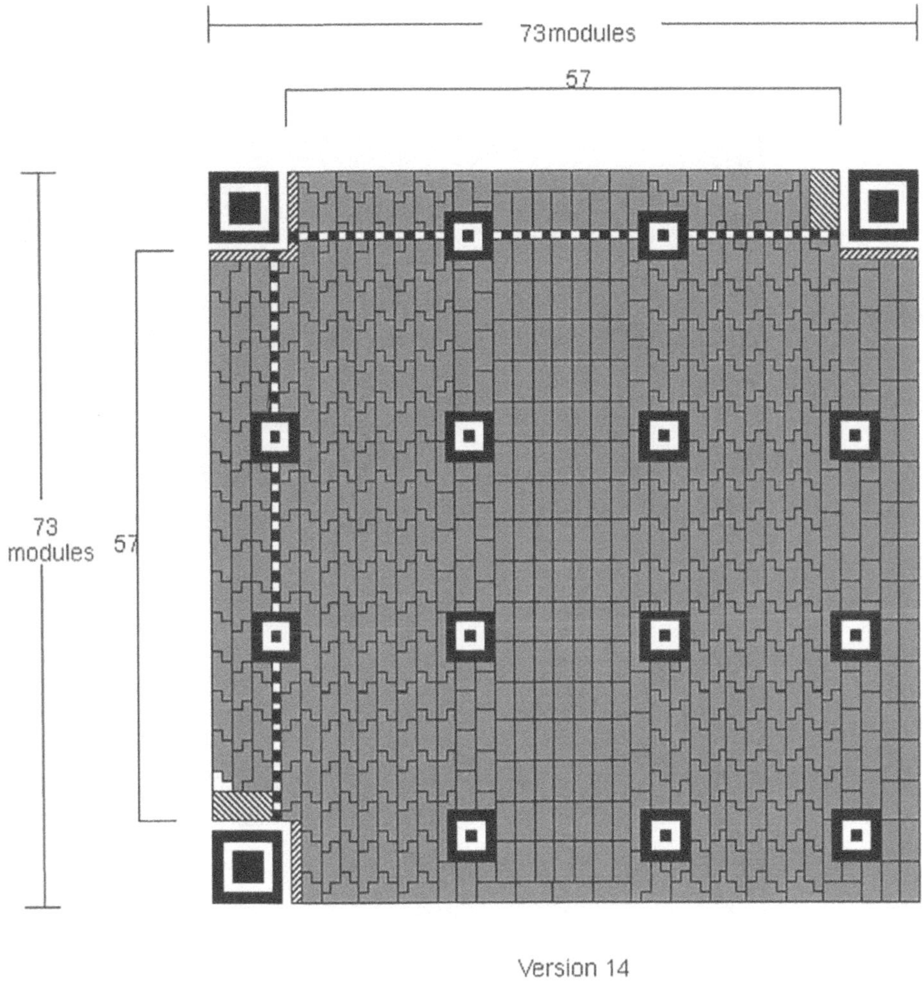

Figure 8 — Version 14 symbol

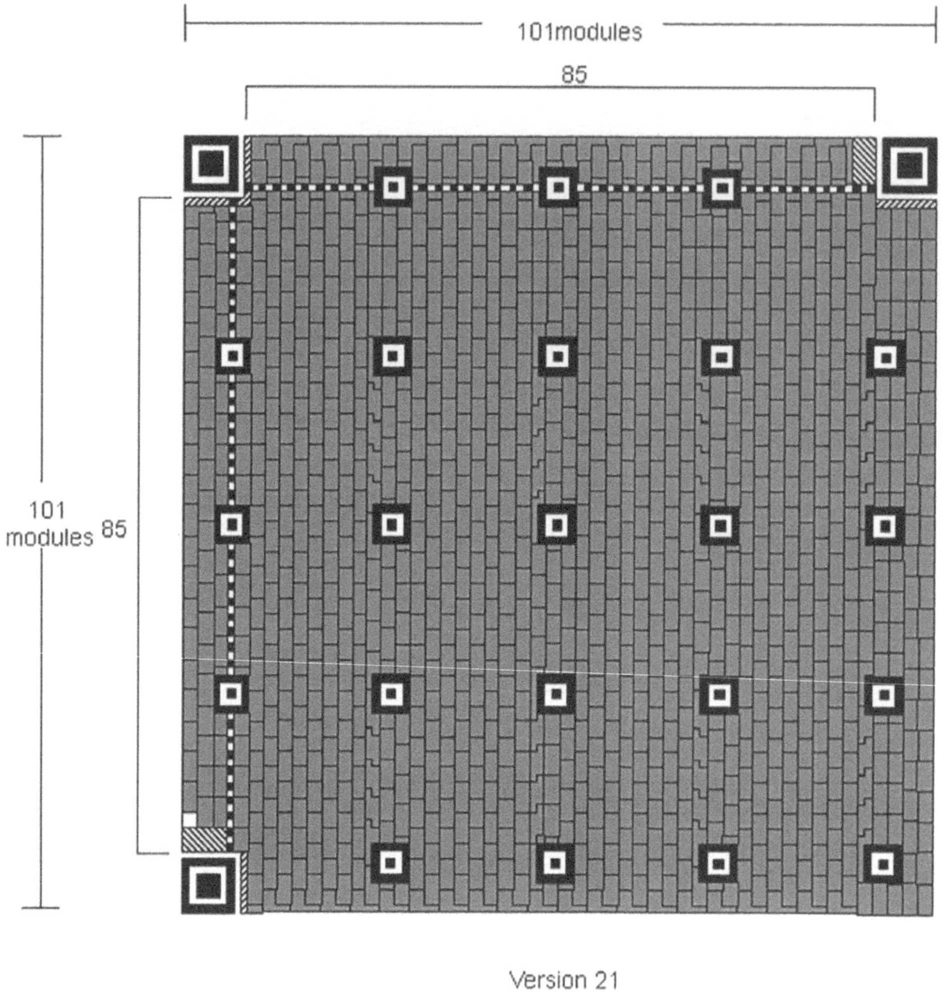

Figure 9 — Version 21 symbol

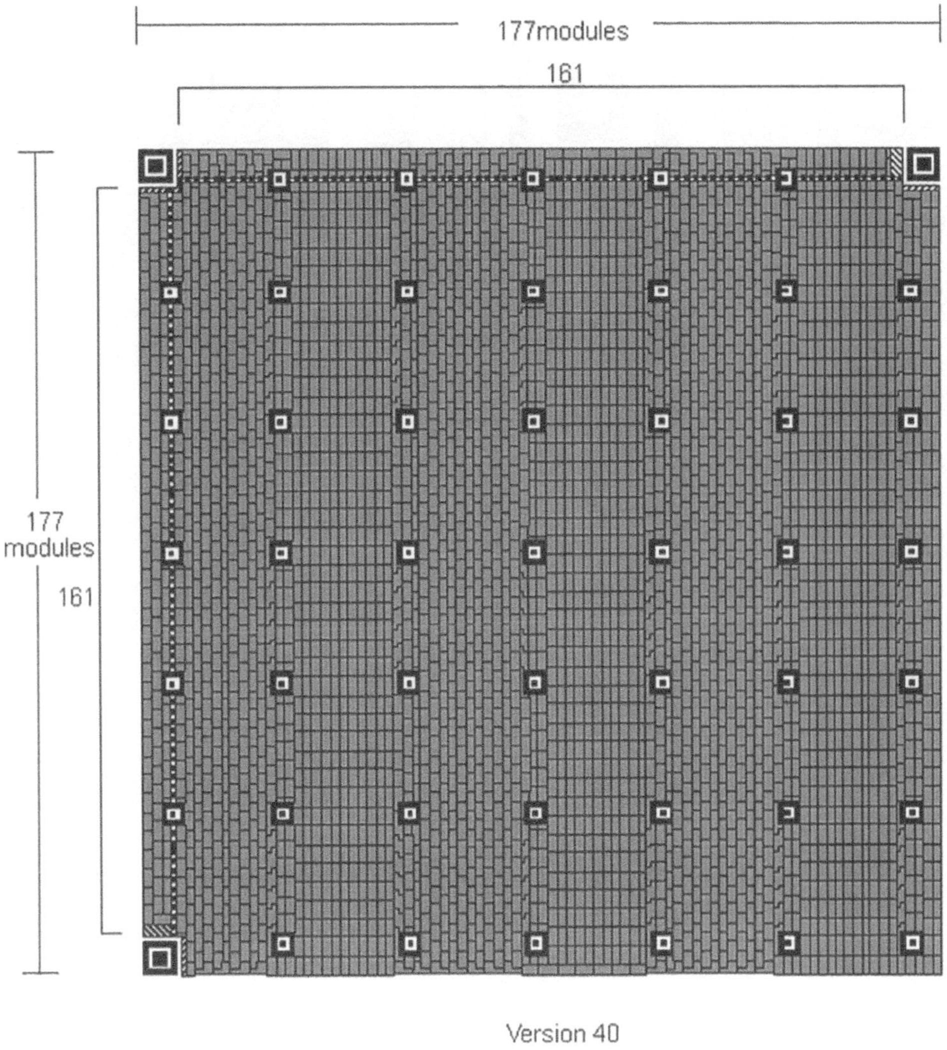

Figure 10 — Version 40 symbol

6.3.2.2 Micro QR Code symbols

There are four sizes of Micro QR Code symbol, referred to as Versions M1 to M4. Version M1 measures 11 × 11 modules, Version M2 13 × 13 modules, Version M3 15 × 15 modules, and Version M4 17 × 17 modules, i.e. increasing in steps of 2 modules per side. Figure 11 illustrates the structure of Micro QR Code Versions M1 to M4.

NOTE Two formats of M3 symbol are shown, which differ only in the codeword placement according to the error correction level.

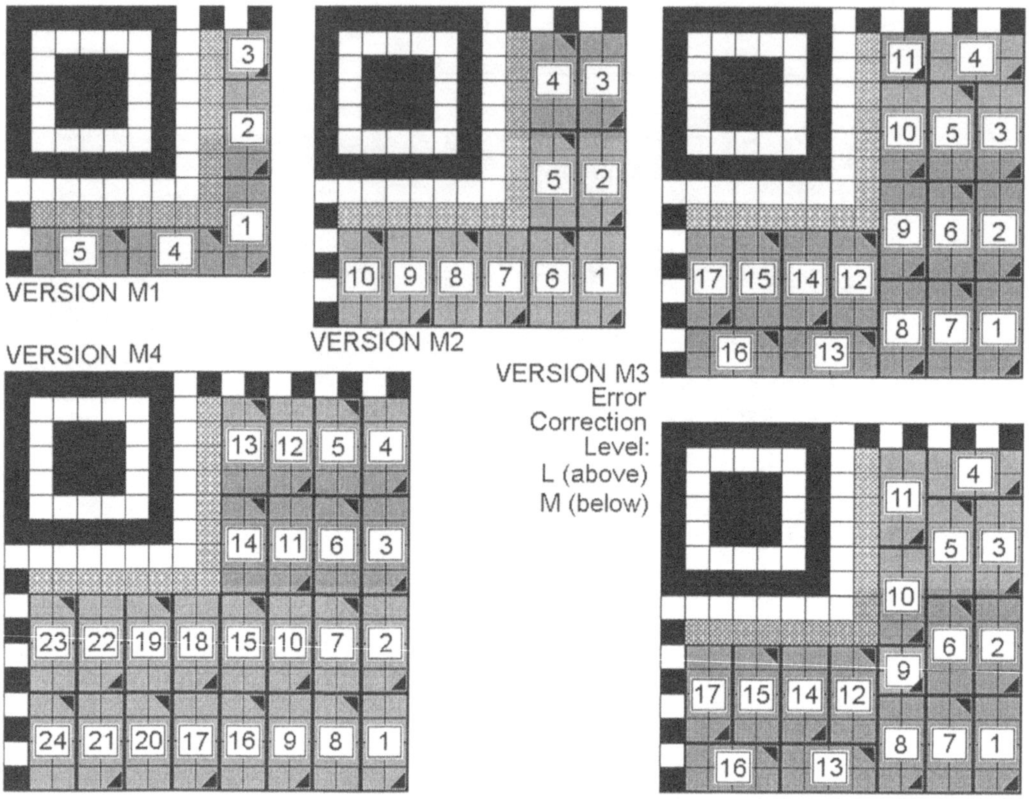

Figure 11 — Versions of Micro QR Code symbol

6.3.3 Finder pattern

6.3.3.1 QR Code symbols

There are three identical Finder Patterns located at the upper left, upper right and lower left corners of the symbol respectively as illustrated in Figure 3. Each finder pattern may be viewed as three superimposed concentric squares and is constructed of dark 7 × 7 modules, light 5 × 5 modules and dark 3 × 3 modules. The ratio of module widths in each finder pattern is 1 : 1 : 3 : 1 : 1 as illustrated in Figure 12. The symbol is preferentially encoded so that similar patterns have a low probability of being encountered elsewhere in the symbol, enabling rapid identification of a possible QR Code symbol in the field of view. Identification of the three finder patterns comprising the Finder Pattern then unambiguously defines the location and rotational orientation of the symbol in the field of view.

6.3.3.2 Micro QR Code symbols

A single finder pattern, as defined in 6.3.3.1, is located at the upper left corner of the symbol as illustrated in Figure 4. Identification of the finder pattern together with the timing patterns unambiguously defines the size, location and rotational orientation of the symbol in the field of view.

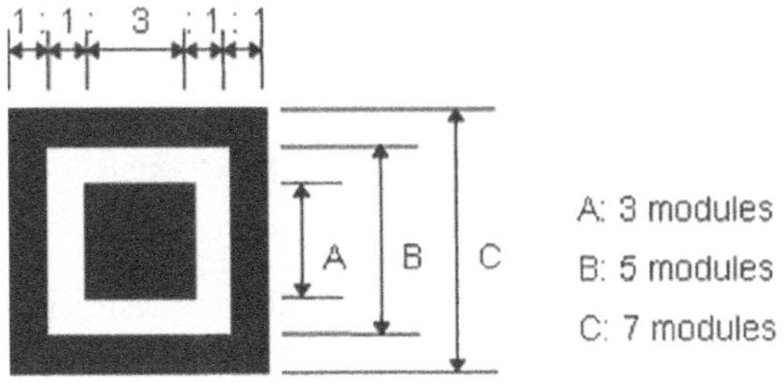

Figure 12 — Structure of finder pattern

6.3.4 Separator

A one-module wide separator, constructed of all light modules, is placed between each finder pattern and the Encoding Region, as illustrated in Figures 3 and 4.

6.3.5 Timing pattern

The horizontal and vertical timing patterns respectively consist of a one module wide row or column of alternating dark and light modules, commencing and ending with a dark module. They enable the symbol density and version to be determined and provide datum positions for determining module coordinates.

In QR Code symbols, the horizontal timing pattern runs across row 6 of the symbol between the separators for the upper finder patterns; the vertical timing pattern similarly runs down column 6 of the symbol between the separators for the left-hand finder patterns. See Figure 3.

In Micro QR Code symbols, the horizontal timing pattern runs across row 0 of the symbol on the right side of the separator to the right hand edge of the symbol; the vertical Timing Pattern similarly runs down column 0 of the symbol below the separator to the bottom edge of the symbol. See Figure 4.

6.3.6 Alignment patterns

Alignment patterns are present only in QR Code symbols of version 2 or larger. Each alignment pattern may be viewed as three superimposed concentric squares and is constructed of dark 5 × 5 modules, light 3 × 3 modules and a single central dark module. The number of alignment patterns depends on the symbol version and they shall be placed in all symbols of Version 2 or larger in positions defined in Annex E.

6.3.7 Encoding region

This region shall contain the symbol characters representing data, those representing error correction codewords, the format information and, where appropriate, the version information. Refer to 7.7.1 for details of the symbol characters. Refer to 7.9 for details of the format information. Refer to 7.10 for details of the version information.

6.3.8 Quiet zone

This is a region which shall be free of all other markings, surrounding the symbol on all four sides. Its nominal reflectance value shall be equal to that of the light modules.

For QR Code symbols its width shall be 4X.

For Micro QR Code symbols its width shall be 2X.

7 Requirements

7.1 Encode procedure overview

This section provides an overview of the steps required to convert input data to a QR Code symbol.

Step 1 Data analysis

Analyze the input data stream to identify the variety of different characters to be encoded. The QR Code format (but not the Micro QR Code format) supports the Extended Channel Interpretation feature, enabling data differing from the default character set to be encoded. QR Code includes several modes (see 7.3) to allow different sub-sets of characters to be converted into symbol characters in efficient ways. Switch between modes as necessary in order to achieve the most efficient conversion of data into a binary string. Select the required Error Detection and Correction Level. If the user has not specified the symbol version to be used, select the smallest version that will accommodate the data. A complete list of symbol versions and capacities is shown in Table 1.

Step 2 Data encoding

Convert the data characters into a bit stream in accordance with the rules for the mode in force, as defined in 7.4.2 to 7.4.6, inserting mode indicators as necessary to change modes at the beginning of each new mode segment, and a Terminator at the end of the data sequence. Split the resulting bit stream into 8-bit codewords. Add Pad Characters as necessary to fill the number of data codewords required for the version.

Step 3 Error correction coding

Divide the codeword sequence into the required number of blocks (as defined in Table 9) to enable the error correction algorithms to be processed. Generate the error correction codewords for each block, appending the error correction codewords to the end of the data codeword sequence.

Step 4 Structure final message

Interleave the data and error correction codewords from each block as described in 7.6 (step 3) and add remainder bits as necessary.

Step 5 Module placement in matrix

Place the codeword modules in the matrix together with the finder pattern, separators, timing pattern, and (if required) alignment patterns.

Step 6 Data masking

Apply the data masking patterns in turn to the encoding region of the symbol. Evaluate the results and select the pattern which optimizes the dark/light module balance and minimizes the occurrence of undesirable patterns.

Step 7 Format and version information

Generate the format information and (where applicable) the version information and complete the symbol.

Table 1 — Codeword capacity of all versions of QR Code

Version	No. of Modules/ side (A)	Function pattern modules (B)	Format and version information modules (C)	Data modules except (C) ($D = A^2 - B - C$)	Data capacity [codewords] [a] (E)	Remainder Bits
M1	11	70	15	36	5	0

[a] All codewords are 8 bits in length, except in versions M1 and M3 where the final data codeword is 4 bits in length

Table 1 *(continued)*

Version	No. of Modules/ side (A)	Function pattern modules (B)	Format and version information modules (C)	Data modules except (C) ($D = A^2 - B - C$)	Data capacity [codewords] [a] (E)	Remainder Bits
M2	13	74	15	80	10	0
M3	15	78	15	132	17	0
M4	17	82	15	192	24	0
1	21	202	31	208	26	0
2	25	235	31	359	44	7
3	29	243	31	567	70	7
4	33	251	31	807	100	7
5	37	259	31	1 079	134	7
6	41	267	31	1 383	172	7
7	45	390	67	1 568	196	0
8	49	398	67	1 936	242	0
9	53	406	67	2 336	292	0
10	57	414	67	2 768	346	0
11	61	422	67	3 232	404	0
12	65	430	67	3 728	466	0
13	69	438	67	4 256	532	0
14	73	611	67	4 651	581	3
15	77	619	67	5 243	655	3
16	81	627	67	5 867	733	3
17	85	635	67	6 523	815	3
18	89	643	67	7 211	901	3
19	93	651	67	7 931	991	3
20	97	659	67	8 683	1 085	3
21	101	882	67	9 252	1 156	4
22	105	890	67	10 068	1 258	4
23	109	898	67	10 916	1 364	4
24	113	906	67	11 796	1 474	4
25	117	914	67	12 708	1 588	4
26	121	922	67	13 652	1 706	4
27	125	930	67	14 628	1 828	4
28	129	1 203	67	15 371	1 921	3
29	133	1 211	67	16 411	2 051	3
30	137	1 219	67	17 483	2 185	3
31	141	1 227	67	18 587	2 323	3
32	145	1 235	67	19 723	2 465	3
33	149	1 243	67	20 891	2 611	3
34	153	1 251	67	22 091	2 761	3
35	157	1 574	67	23 008	2 876	0

[a] All codewords are 8 bits in length, except in versions M1 and M3 where the final data codeword is 4 bits in length

Table 1 *(continued)*

Version	No. of Modules/side (A)	Function pattern modules (B)	Format and version information modules (C)	Data modules except (C) (D = A^2 – B – C)	Data capacity [codewords] [a] (E)	Remainder Bits
36	161	1 582	67	24 272	3 034	0
37	165	1 590	67	25 568	3 196	0
38	169	1 598	67	26 896	3 362	0
39	173	1 606	67	28 256	3 532	0
40	177	1 614	67	29 648	3 706	0

[a] All codewords are 8 bits in length, except in versions M1 and M3 where the final data codeword is 4 bits in length

7.2 Data analysis

Analyze the input data string to determine its content and select the default or other appropriate ECI and the appropriate mode to encode each sequence as described in 7.4. Each mode in sequence from Numeric mode to Kanji mode progressively requires more bits per character. It is possible to switch from mode to mode within a symbol in order to minimize the bit stream length for data, parts of which can more efficiently be encoded in one mode than other parts, e.g. numeric sequences followed by alphanumeric sequences. It is in theory most efficient to encode data in the mode requiring the fewest bits per data character, but as there is some overhead in the form of mode indicator and character count indicator associated with each mode change, it may not always result in the shortest overall bit stream to change modes for a small number of characters. Also, because the capacity of symbols increases in discrete steps from one version to the next, it may not always be necessary to achieve the maximum conversion efficiency in every case. Guidance on minimising the bit stream length is given in Annex J. In Micro QR Code symbols, there are restrictions on the modes available in the smaller versions. Annex J.2 shows the Micro QR Code symbol versions appropriate for various combinations of two modes.

7.3 Modes

7.3.1 General

The modes defined below are based on the character values and assignments associated with the default ECI. When any other ECI is in force (in QR Code symbols only), the byte values rather than the specific character assignments shall be used to select the optimum data compaction mode. For example, Numeric mode would be appropriate if there is a sequence of data byte values within the range 30_{HEX} to 39_{HEX} inclusive. In this case the compaction is carried out using the default numeric or alphabetic equivalents of the byte values.

7.3.2 Extended Channel Interpretation (ECI) mode

The Extended Channel Interpretation (ECI) protocol defined in the AIM Inc. International Technical Specification *Extended Channel Interpretations*, allows the output data stream to have interpretations different from that of the default character set. The ECI protocol is defined consistently across a number of symbologies. The ECI protocol provides a consistent method to specify particular interpretations of byte values before printing and after decoding. The ECI protocol is not supported in Micro QR Code symbols.

The default interpretation for QR Code is ECI **000003** representing the ISO/IEC 8859-1 character set.

International applications using other character sets should use the ECI protocol. For instance, the interpretation corresponding to the JIS8 and Shift JIS character sets is ECI **000020**.

The effect of ECI mode is to insert an ECI escape sequence at that point in the data. It is immediately followed by another mode indicator (e.g. for efficient data encoding) and remains in force until the end of the message or a subsequent ECI mode indicator.

7.3.3 Numeric mode

Numeric mode encodes data from the decimal digit set (**0 - 9**) (byte values 30_{HEX} to 39_{HEX}). Normally, 3 data characters are represented by 10 bits.

7.3.4 Alphanumeric mode

Alphanumeric mode encodes data from a set of 45 characters, i.e. 10 numeric digits (**0 - 9**) (byte values 30_{HEX} to 39_{HEX}), 26 alphabetic characters (**A - Z**) (byte values 41_{HEX} to $5A_{HEX}$) , and 9 symbols (**SP, $, %, *, +, -, ., /, :**) (byte values 20_{HEX}, 24_{HEX}, 25_{HEX}, $2A_{HEX}$, $2B_{HEX}$, 2D to $2F_{HEX}$, $3A_{HEX}$ respectively). Normally, two input characters are represented by 11 bits.

Alphanumeric mode is not available in Version M1 Micro QR Code symbols.

7.3.5 Byte mode

In this mode, data is encoded at 8 bits per character.

In closed-system national or application-specific implementations of QR Code, an alternative 8-bit character set, for example as defined in an appropriate part of ISO/IEC 8859, may be specified for Byte mode. When an alternative character set is specified, however, the parties intending to read the QR Code symbols require to be notified of the applicable character set in the application specification or by bilateral agreement.

Byte mode is not available in Version M1 or M2 Micro QR Code symbols.

7.3.6 Kanji mode

The Kanji mode efficiently encodes Kanji characters in accordance with the Shift JIS system based on JIS X 0208. The Shift JIS values are shifted from the JIS X 0208 values. JIS X 0208 gives details of the shift coded representation. Each two-byte character value is compacted to a 13-bit binary codeword.

When the character set specified for 8-bit byte mode makes use of byte values in the ranges 81_{HEX} to $9F_{HEX}$ and/or $E0_{HEX}$ to EB_{HEX}, it may not be possible to use Kanji mode unambiguously, as reading systems will be unable to determine from the transmitted data whether such byte values are the lead byte of a double byte character. It may be possible to achieve a shorter bit stream by using the Kanji mode compaction rules when an appropriate sequence of byte values occurs in the data (i.e. lead bytes in the ranges 81_{HEX} to $9F_{HEX}$ and/or $E0_{HEX}$ to EB_{HEX} followed by trailer bytes in the range 40_{HEX} to FC_{HEX},except $7F_{HEX}$, or EB_{HEX} followed by 40_{HEX} to BF_{HEX}). Figure H.1 shows the byte combinations graphically.

Kanji mode is not available in version M1 or M2 Micro QR Code symbols.

7.3.7 Mixing modes

The QR Code symbol may contain sequences of data in a combination of any of the modes described in 7.3.2 to 7.3.9. Micro QR Code symbols may contain sequences of data in a combination of any of the modes available for the version of the symbol and described in 7.3.3 to 7.3.7.

Refer to Annex J for guidance on selecting the most efficient way of representing a given input data string in multiple modes in QR Code symbols, and to Annex J.3 for the available versions of Micro QR Code symbols for given combinations of data in two modes.

7.3.8 Structured Append mode

Structured Append mode is used to split the encoding of the data from a message over a number of QR Code symbols. All of the symbols require to be read and the data message can be reconstructed in the correct sequence. The Structured Append header is encoded in each symbol to identify the length of the sequence and the symbol's position in it, and verify that all the symbols read belong to the same message. Refer to 8 for details of encoding in Structured Append mode.

Structured Append mode is not available for Micro QR Code symbols.

7.3.9 FNC1 mode

FNC1 mode is used for messages containing specific data formats. In the "1st position" it designates data formatted in accordance with the GS1 General Specifications. In the "2nd position" it designates data formatted in accordance with a specific industry application previously agreed with AIM Inc. FNC1 mode applies to the entire symbol and is not affected by subsequent mode indicators.

NOTE "1st position" and 2nd position" do not refer to actual locations but are based on the positions of the character in Code 128 symbols, when used in an equivalent manner.

FNC1 mode is not available for Micro QR Code symbols.

7.4 Data encoding

7.4.1 Sequence of data

Input data is converted into a bit stream consisting of one or more segments each in a separate mode. In the default ECI, the bit stream commences with the first mode indicator. If the initial ECI is other than the default ECI, the bit stream commences with an ECI header, followed by the first segment.

The ECI header (if present) shall comprise:

— ECI mode indicator (4 bits)

— ECI Designator (8, 16 or 24 bits)

The ECI header shall begin with the first (most significant) bit of the ECI mode indicator and end with the final (least significant) bit of the ECI Designator.

The remainder of the bit stream is then made up of segments each comprising:

— Mode indicator

— Character count indicator

— Data bit stream

Each mode segment shall begin with the first (most significant) bit of the mode indicator and end with the final (least significant) bit of the data bit stream. There shall be no explicit separator between segments as their length is defined unambiguously by the rules for the mode in force and the number of input data characters.

To encode a sequence of input data in a given mode, the steps defined in sections 7.4.2 to 7.4.7 shall be followed. Table 2 defines the mode indicators for each mode. Table 3 defines the length of the character count indicator, which varies according to the mode and the symbol version in use.

Table 2 — Mode indicators for QR Code

Mode	QR Code symbols	Micro QR Code symbols			
Version	**all**	**M1**	**M2**	**M3**	**M4**
Mode indicator length (bits)	4	0	1	2	3
ECI	**0111**	n/a	n/a	n/a	n/a
Numeric	**0001**	n/a	0	00	000
Alphanumeric	**0010**	n/a	1	01	001
Byte	**0100**	n/a	n/a	10	010
Kanji	**1000**	n/a	n/a	11	011
Structured Append	**0011**	n/a	n/a	n/a	n/a
FNC1 [a]	0101 (1st position) **1001** (2nd position)	n/a	n/a	n/a	n/a
Terminator (End of Message) [b]	**0000**	**000**	**00000**	**0000000**	**000000000**

[a] See 7.4.8.2 and 7.4.8.3.

[b] The Terminator is not a mode indicator as such.

Table 3 — Number of bits in character count indicator for QR Code

Version	Numeric mode	Alphanumeric mode	Byte mode	Kanji mode
M1	**3**	n/a	n/a	n/a
M2	**4**	**3**	n/a	n/a
M3	**5**	**4**	**4**	**3**
M4	**6**	**5**	**5**	**4**
1 to 9	**10**	**9**	**8**	**8**
10 to 26	**12**	**11**	**16**	**10**
27 to 40	**14**	**13**	**16**	**12**

The end of the data in the complete symbol is indicated by a Terminator consisting of between 3 and 9 zero bits (see Table 2), which is omitted or abbreviated if the remaining symbol capacity after the data bit stream is less than the required bit length of Terminator. The Terminator is not a mode indicator as such.

7.4.2 Extended Channel Interpretation (ECI) mode

7.4.2.1 General

This mode, used for encoding data subject to alternative interpretations of byte values (e.g. alternative character sets) in accordance with the AIM ECI specification which defines the pre-processing of this type of data, is invoked by the use of mode indicator **0111.**

The Extended Channel Interpretation can only be used with readers enabled to transmit the Symbology Identifier. Readers that cannot transmit the Symbology Identifier cannot transmit the data from any symbol containing an ECI.

Input ECI data shall be handled by the encoding system as a series of byte values.

Data in an ECI sequence may be encoded in whatever mode or modes permit the most efficient encoding of the byte values of the data, irrespective of their significance. For example, a sequence of bytes in the range 30_{HEX} to 39_{HEX} could be encoded in Numeric mode (see 7.4.3) as though it were a sequence of digits 0 – 9 even though it might not actually represent numeric data. In order to determine the value of the character count indicator, the number of bytes (or, in Kanji mode, of byte pairs) shall be used.

7.4.2.2 ECI Designator

Each Extended Channel Interpretation is designated by a six-digit assignment number which is encoded in the QR Code symbol as the first one, two or three codewords following the ECI mode indicator. The encoding rules are defined in Table 4. The ECI Designator appears in the data to be encoded as character $5C_{HEX}$ [\ or backslash (reverse solidus) in ISO/IEC 8859-1, ¥ or yen sign in JIS8] followed by the six digit assignment number. Where $5C_{HEX}$ appears as true data it shall be doubled in the data string before encoding in symbols to which the ECI protocol applies.

When a single occurrence of $5C_{HEX}$ is encountered in the input to the decoder, an ECI mode indicator is inserted followed by the ECI Designator. When a doubled $5C_{HEX}$ is encountered, it is encoded as two $5C_{HEX}$ bytes.

On decoding, the binary pattern of the first ECI Designator codeword (i.e. the codeword following the mode indicator in ECI mode), determines the length of the ECI Designator sequence. The number of **1** bits before the first **0** bit defines the number of additional codewords after the first used to represent the ECI Assignment number. The bit sequence after the first **0** bit is the binary representation of the ECI Assignment number. The lower numbered ECI assignments may be encoded in multiple ways, but the shortest way is preferred.

Table 4 — Encoding ECI Assignment Number

ECI Assignment Value	No. of Codewords	Codeword values
000000 to 000127	1	**0**bbbbbbb
000000 to 016383	2	**10**bbbbbb bbbbbbbb
000000 to 999999	3	**110**bbbbb bbbbbbbb bbbbbbbb
		where b ... b is the binary value of the ECI Assignment number

Example

Assume data to be encoded is in Greek, using character set ISO/IEC 8859-7 (ECI 000009) in version 1-H symbol.

Data to be encoded: \000009**ΑΒΓΔΕ** (character values $A1_{HEX}$, $A2_{HEX}$, $A3_{HEX}$, $A4_{HEX}$, $A5_{HEX}$)

Bit sequence in symbol:

ECI mode indicator **0111**

ECI Assignment number (000009) **0 0001001**

Mode indicator (byte) **0100**

Character count indicator (5) **00000101**

Data: **10100001 10100010 10100011 10100100 10100101**

Final bit string: **0111 00001001 0100 00000101 10100001 10100010 10100011 10100100 10100101**

See 14.3 for example of transmission of this data following decoding.

7.4.2.3 Multiple ECIs

Refer to the AIM ECI specification for the rules defining the effect of a subsequent ECI Designator in an ECI data segment. For example, data to which a character set ECI has been applied may also be subject to encryption or compaction using a transformation ECI which will co-exist with the initial ECI, or a second character set ECI will have the effect of terminating the first ECI and starting a new ECI segment. Where any ECI Designator appears in the data, it shall be encoded in the QR Code symbol in accordance with 7.4.2.2 and shall commence a new mode segment.

7.4.2.4 ECIs and Structured Append

Any ECI(s) invoked shall apply subject to the rules defined above and in the AIM ECI specification until the end of the encoded data or a change of ECI (signalled by mode indicator **0111**). If the encoded data in the ECI(s) extends through two or more symbols in Structured Append mode, it is necessary to provide an ECI header consisting of ECI mode indicator and ECI Designator number for each ECI in force, immediately following the Structured Append header, in subsequent symbols in which the ECI continues in force.

7.4.3 Numeric mode

The input data string is divided into groups of three digits, and each group is converted to its 10-bit binary equivalent. If the number of input digits is not an exact multiple of three, the final one or two digits are converted to 4 or 7 bits respectively. The binary data is then concatenated and prefixed with the mode indicator and the character count indicator. The mode indicator in the Numeric mode has either 4 bits for QR Code symbols or the number of bits defined in Table 2 for Micro QR Code symbols, and the character count indicator has the number of bits defined in Table 3. The number of input data characters is converted to its binary equivalent and added as the character count indicator after the mode indicator and before the binary data sequence.

EXAMPLE 1 (for Version 1-H symbol)

Input data: **01234567**

1. Divide into groups of three digits: **012 345 67**
2. Convert each group to its binary equivalent: **012 → 0000001100**
   345 → 0101011001
   67 → 1000011
3. Connect the binary data in sequence: **0000001100 0101011001 1000011**
4. Convert character count indicator to binary (10 bits for version 1-H):

   No. of input data characters: **8 → 0000001000**

5. Add mode indicator **0001** and character count indicator to binary data:

   0001 0000001000 0000001100 0101011001 1000011

EXAMPLE 2 (for Micro QR Code version M3-M symbol)

Input data: **0123456789012345**

1. Divide into groups of three digits: **012 345 678 901 234 5**
2. Convert each group to its binary equivalent: **012 = 0000001100**
   345 = 0101011001

678 = 1010100110

901 = 1110000101

234 = 0011101010

5 = 0101

3. Connect the binary data in sequence:

0000001100 0101011001 1010100110 1110000101 0011101010 0101

4. Convert character count indicator to binary (5 bits for version M3-M):

No. of input data characters: **16 = 10000**

5. Add mode indicator (**00** for version M3-M) and character count indicator to binary data:

00 10000 0000001100 0101011001 1010100110 1110000101 0011101010 0101

For any number of data characters the length of the bit stream in Numeric mode is given by the following formula:

$$B = M + C + 10(D \text{DIV} 3) + R$$

where:

B number of bits in bit stream

M number of bits in mode indicator (4 for QR Code symbols, or as shown in Table 2 for Micro QR Code symbols)

C number of bits in character count indicator (from Table 3)

D number of input data characters

R 0 if (*D* MOD 3) = 0

R 4 if (*D* MOD 3) = 1

R 7 if (*D* MOD 3) = 2

7.4.4 Alphanumeric mode

Each input data character is assigned a character value V from 0 to 44 according to Table 5.

Table 5 — Encoding/decoding table for Alphanumeric mode

Char.	Value	Char.	Value	Char.	Value	Char.	Value	Char.	Value	Char.	Value	Char.	Value	Char.	Value
0	0	6	6	C	12	I	18	O	24	U	30	SP	36	.	42
1	1	7	7	D	13	J	19	P	25	V	31	$	37	/	43
2	2	8	8	E	14	K	20	Q	26	W	32	%	38	:	44
3	3	9	9	F	15	L	21	R	27	X	33	*	39		
4	4	A	10	G	16	M	22	S	28	Y	34	+	40		
5	5	B	11	H	17	N	23	T	29	Z	35	-	41		

Input data characters are divided into groups of two characters which are encoded as 11-bit binary codes. The character value of the first character is multiplied by 45 and the character value of the second digit is added to the product. The sum is then converted to an 11-bit binary number. If the number of input data characters is not a multiple of two, the character value of the final character is encoded as a

6-bit binary number. The binary data is then concatenated and prefixed with the mode indicator and the character count indicator. The mode indicator in the Alphanumeric mode has either 4 bits for QR Code symbols or the number of bits defined in Table 2 for Micro QR Code symbols, and the character count indicator has the number of bits defined in Table 3. The number of input data characters is converted to its binary equivalent and added as the character count indicator after the mode indicator and before the binary data sequence.

In FNC1 mode symbols the **FNC1** character may occur in the data. It is represented in Alphanumeric mode by the character **%**. Refer to 7.4.8.2, 7.4.8.3 and 14.4 for details of the encoding and transmission of **FNC1** and **%**.

EXAMPLE (for Version 1-H symbol)

Input data:	**AC-42**
1. Determine character values according to Table 5.	**AC-42 → (10,12,41,4,2)**
2. Divide the result into groups of two decimal values:	**(10,12) (41,4) (2)**
3. Convert each group to its 11-bit binary equivalent:	**(10,12) 10*45+12 → 462 → 00111001110**
	(41,4) 41*45+4 → 1849 → 11100111001
	(2) → 2 → 000010
4. Connect the binary data in sequence:	**00111001110 11100111001 000010**

5. Convert character count indicator to binary (9 bits for version 1-H):

No. of input data characters: **5 → 000000101**

6. Add mode indicator **0010** and character count indicator to binary data:

0010 000000101 00111001110 11100111001 000010

For any number of data characters the length of the bit stream in Alphanumeric mode is given by the following formula:

$$B = M + C + 11(D\,\mathrm{DIV}\,2) + 6(D\,\mathrm{MOD}\,2)$$

where:

- B number of bits in bit stream
- M number of bits in mode indicator (4 for QR Code symbols, or as shown in Table 2 for Micro QR Code symbols)
- C number of bits in character count indicator (from Table 3)
- D number of input data characters

7.4.5 Byte mode

In this mode, one 8-bit codeword directly represents the byte value of the input data character, i.e. a density of 8 bits/character.

Table 6 — Encoding/decoding table for ISO/IEC 8859-1 character set

Byte	Char.	Byte	Char.	Byte	Char.	Byte	Char.	Byte	Char.	Byte	Char.	Byte	Char.	Byte	Char.
0	NUL	32	space	64	@	96	`	128		160	NBSP	192	À	224	à

Table 6 *(continued)*

Byte	Char.	Byte	Char.	Byte	Char.	Byte	Char.	Byte	Char.	Byte	Char.	Byte	Char.	Byte	Char.
1	SOH	33	!	65	A	97	a	129		161	¡	193	Á	225	á
2	STX	34	"	66	B	98	b	130		162	¢	194	Â	226	â
3	ETX	35	#	67	C	99	c	131		163	£	195	Ã	227	ã
4	EOT	36	$	68	D	100	d	132		164	¤	196	Ä	228	ä
5	ENQ	37	%	69	E	101	e	133		165	¥	197	Å	229	å
6	ACK	38	&	70	F	102	f	134		166	¦	198	Æ	230	æ
7	BEL	39	'	71	G	103	g	135		167	§	199	Ç	231	ç
8	BS	40	(	72	H	104	h	136		168	¨	200	È	232	è
9	HT	41	)	73	I	105	I	137		169	©	201	É	233	é
10	LF	42	*	74	J	106	j	138		170	ª	202	Ê	234	ê
11	VT	43	+	75	K	107	k	139		171	«	203	Ë	235	ë
12	FF	44	,	76	L	108	l	140		172	¬	204	Ì	236	ì
13	CR	45	-	77	M	109	m	141		173	SHY	205	Í	237	í
14	SO	46	.	78	N	110	n	142		174	®	206	Î	238	î
15	SI	47	/	79	O	111	o	143		175	¯	207	Ï	239	ï
16	DLE	48	0	80	P	112	p	144		176	°	208	Ð	240	ð
17	DC1	49	1	81	Q	113	q	145		177	±	209	Ñ	241	ñ
18	DC2	50	2	82	R	114	r	146		178	²	210	Ò	242	ò
19	DC3	51	3	83	S	115	s	147		179	³	211	Ó	243	ó
20	DC4	52	4	84	T	116	t	148		180	´	212	Ô	244	ô
21	NAK	53	5	85	U	117	u	149		181	µ	213	Õ	245	õ
22	SYN	54	6	86	V	118	v	150		182	¶	214	Ö	246	ö
23	ETB	55	7	87	W	119	w	151		183	·	215	×	247	÷
24	CAN	56	8	88	X	120	x	152		184	¸	216	Ø	248	ø
25	EM	57	9	89	Y	121	y	153		185	¹	217	Ù	249	ù
26	SUB	58	:	90	Z	122	z	154		186	º	218	Ú	250	ú
27	ESC	59	;	91	[	123	{	155		187	»	219	Û	251	û
28	FS	60	<	92	\	124	|	156		188	¼	220	Ü	252	ü
29	GS	61	=	93	]	125	}	157		189	½	221	Ý	253	ý
30	RS	62	>	94	^	126	~	158		190	¾	222	Þ	254	þ
31	US	63	?	95	_	127	DEL	159		191	¿	223	ß	255	ÿ

NOTE 1 In the JIS8 character set (see Table H.1), byte values 80_{HEX} to $9F_{HEX}$ and $E0_{HEX}$ to FF_{HEX} are not assigned but are reserved values. Some of those values are used as the first byte in the Shift JIS character set (see Table H.2) and may be used to distinguish between the JIS8 and Shift JIS character sets, or to enable Kanji mode compaction to be carried out. JIS X 0208 gives details of the shift coded representation.

NOTE 2 Byte values 00_{HEX} to $7F_{HEX}$ in the JIS8 character set correspond to ISO/IEC 8859-1 and ISO/IEC 646 IRV, except values $5C_{HEX}$ and $7E_{HEX}$.

The binary data is then concatenated and prefixed with the mode indicator and the character count indicator. The mode indicator in the Byte mode has either 4 bits for QR Code symbols or the number of bits defined in Table 2 for Micro QR Code symbols, and the character count indicator has the number of bits defined in Table 3. The number of input data characters is converted to its binary equivalent and added after the mode indicator and before the binary data sequence.

For any number of data characters the length of the bit stream in Byte mode is given by the following formula:

$$B = M + C + 8D$$

where:

B number of bits in bit stream

M number of bits in mode indicator (4 for QR Code symbols, or as shown in Table 2 for Micro QR Code symbols)

C number of bits in character count indicator (from Table 3)

D number of input data characters

7.4.6 Kanji mode

In the Shift JIS system, Kanji characters are represented by a two byte combination. These byte values are shifted from the JIS X 0208 values. JIS X 0208 gives details of the shift coded representation. Input data characters in Kanji mode are compacted to 13-bit binary codewords as defined below. The binary data is then concatenated and prefixed with the mode indicator and the character count indicator. The mode indicator in the Numeric mode has either 4 bits for QR Code symbols or the number of bits defined in Table 2 for Micro QR Code symbols, and the character count indicator has the number of bits defined in Table 3. The number of input data characters is converted to its binary equivalent and added as the character count indicator after the mode indicator and before the binary data sequence.

1. For characters with Shift JIS values from 8140_{HEX} to $9FFC_{HEX}$:
   a) Subtract 8140_{HEX} from Shift JIS value;
   b) Multiply most significant byte of result by $C0_{HEX}$;
   c) Add least significant byte to product from b);
   d) Convert result to a 13-bit binary string.
2. For characters with Shift JIS values from $E040_{HEX}$ to $EBBF_{HEX}$:
   a) Subtract $C140_{HEX}$ from Shift JIS value;
   b) Multiply most significant byte of result by $C0_{HEX}$;
   c) Add least significant byte to product from b);
   d) Convert result to a 13-bit binary string.

EXAMPLES:

Input character	"点"	"茗"
(Shift JIS value):	935F	E4AA
1. Subtract 8140 or C140	935F - 8140 = 121F	E4AA - C140 = 236A
2. Multiply m.s.b. by C0	12 × C0 = D80	23 × C0 = 1A40
3. Add l.s.b.	D80 + 1F = D9F	1A40 + 6A = 1AAA
4. Convert to 13-bit binary	0D9F →0 1101 1001 1111	1AAA →1 1010 1010 1010

3. For all characters:

   e) Prefix binary sequence representing input data characters with mode indicator (from Table 2) and character count indicator binary equivalent (number of bits defined in Tables);

For any number of data characters the length of the bit stream in Kanji mode is given by the following formula:

$$B = M + C + 13D$$

where:

B number of bits in bit stream

M number of bits in mode indicator (4 for QR Code symbols, or as shown in Table 2 for Micro QR Code symbols)

C number of bits in character count indicator (from Table 3)

D number of input data characters

7.4.7 Mixing modes

There is the option for a symbol to contain sequences of data in one mode and then to change modes if the data content requires it, or in order to increase the density of encoding. Refer to Annex J for guidance. Each segment of data is encoded in the appropriate mode as indicated in 7.4.2 to 7.4.6, with the basic structure mode indicator/character count indicator/Data and followed immediately by the mode indicator commencing the next segment. Figure 13 illustrates the structure of data containing *n* segments.

Segment 1			Segment 2			……	Segment *n*			
mode indicator 1	character count indicator	Data	mode indicator 2	character count indicator	Data	……	mode indicator *n*	character count indicator	Data	Terminator

Figure 13 — Format of mixed mode data

7.4.8 FNC1 modes

7.4.8.1 General

In QR Code symbols, there are two mode indicators which are used cumulatively with those defined in 7.3.2 to 7.3.9 and 7.4.2 to 7.4.7 to identify symbols encoding messages formatted according to specific predefined industry or application specifications. These (together with any associated parameter data) precede the mode indicator(s) used to encode the data efficiently. When these mode indicators are used, it is necessary for the decoder to transmit the Symbology Identifier as defined in 14.2 and Annex F.

7.4.8.2 FNC1 in first position

NOTE "first position" is not used in a literal sense but is a historical reference to the position of the FNC1 symbol character in Code 128 symbols.

This mode indicator identifies symbols encoding data formatted according to the GS1 Application Identifiers standard. For this purpose, it shall only be used once in a symbol and shall be placed immediately before the first mode indicator used for efficient data encoding (Numeric, Alphanumeric, Byte or Kanji), and after any ECI or Structured Append header. Where the GS1 specifications call for the FNC1 character (in other symbologies which use this special character) to be used as a data field separator (i.e. at the end of a variable-length data field), QR Code symbols shall use the **%** character in

Alphanumeric mode or character **GS** (byte value $1D_{HEX}$) in Byte mode to perform this function. If the **%** character occurs as part of the data it shall be encoded as **%%**. Decoders encountering **%** in these symbols shall transmit it as ASCII/JIS8 value $1D_{HEX}$, and if **%%** is encountered it shall be transmitted as a single **%** character.

EXAMPLE 1

Input data: **0104912345123459** (Application Identifier 01 = GS1 article no., fixed length; data: 04912345123459)

15970331 (Application Identifier 15 = "Best before" date YYMMDD, fixed length; data: 31 March 1997)

30128 (Application Identifier 30 = quantity, variable length; data: 128) (requires separator character)

10ABC123 (Application Identifier 10 = batch number, variable length; data: ABC123)

Data to be encoded:

01049123451234591597033130128%10ABC123

Bit sequence in symbol:

0101 (mode indicator, FNC1 implied in 1st position)

0001 (mode indicator, Numeric mode)

0000011101 (character count indicator, 29)

<data bits for **01049123451234591597033130128**>

0010 (mode indicator, Alphanumeric mode)

000001001 (character count indicator, 9)

<data bits for **%10ABC123**>

Transmitted data (see 14.2 and Annex F)

]Q301049123451234591597033130128**<$1D_{HEX}$>**10ABC123

EXAMPLE 2 Encoding/transmission of % character in data:

Input data:	**123%**
Encoded as:	**123%%**
Transmitted as:	**123%**

7.4.8.3 FNC1 in second position

NOTE "second position" is not used in a literal sense but is a historical reference to the position of the FNC1 symbol character in Code 128 symbols.

This mode indicator identifies symbols formatted in accordance with specific industry or application specifications previously agreed with AIM International. It is immediately followed by a one-byte codeword the value of which is that of the Application Indicator assigned to identify the specification concerned by AIM International. For this purpose, it shall only be used once in a symbol and shall be placed immediately before the first mode indicator used for efficient data encoding (Numeric, Alphanumeric, Byte or Kanji), and after any ECI or structured Append header. An Application Indicator may take the form of any single Latin alphabetic character from the set {a - z, A - Z} (represented by the

ASCII value of the character plus 100) or a two-digit number (represented by its numeric value directly) and shall be transmitted by the decoder as the first one or two characters immediately preceding the data. Where the application specifications call for the FNC1 character (in other symbologies which use this special character) to be used as a data field separator, QR Code symbols shall use the **%** character in Alphanumeric mode or character **GS** (ASCII/JIS8 value $1D_{HEX}$) in Byte mode to perform this function. If the **%** character occurs as part of the data it shall be encoded as **%%**. Decoders encountering **%** in these symbols shall transmit it as ASCII/JIS8 value $1D_{HEX}$, and if **%%** is encountered it shall be transmitted as a single **%** character.

EXAMPLE:

NOTE Application Indicator 37 has not been assigned at the time of publication to any organisation and the data content of the example is purely arbitrary.

Application Indicator: 37

Input data: AA1234BBB112text text text text<CR>

Bit sequence in symbol:

1001 (mode indicator, FNC1 implied in 2nd position)

00100101 (Application Indicator, 37)

0010 (mode indicator, Alphanumeric mode)

000001100 (character count indicator, 12)

<data bits for **AA1234BBB112**>

0100 (mode indicator, Byte mode)

00010100 (character count indicator, 20)

<data bits for **text text text text<CR>** >

Transmitted data:

]**Q5**37AA1234BBB112text text text text<CR>

7.4.9 Terminator

The end of data in the symbol is signalled by the Terminator sequence of **0** bits, as defined in Table 2, appended to the data bit stream following the final mode segment. The Terminator shall be omitted if the data bit stream completely fills the capacity of the symbol, or abbreviated if the remaining capacity of the symbol is less than the required bit length of Terminator.

7.4.10 Bit stream to codeword conversion

The bit streams corresponding to each mode segment shall be connected in order. The Terminator shall be appended to the complete bit stream as defined in 7.4.9. The resulting message bit stream shall then be divided into codewords. All codewords are 8 bits in length, except for the final data symbol character in Micro QR Code versions M1 and M3 symbols, which is 4 bits in length. If the bit stream length is such that it does not end at a codeword boundary, padding bits with binary value 0 shall be added after the final bit (least significant bit) of the data stream to extend it to the codeword boundary. The message bit stream shall then be extended to fill the data capacity of the symbol corresponding to the Version and Error Correction Level, as defined in Table 8, by adding the Pad Codewords **11101100** and **00010001** alternately. For Micro QR Code versions M1 and M3 symbols, the final data codeword is 4 bits long. The Pad Codeword used in the final data symbol character position in Micro QR Code versions M1 and M3 symbols shall be represented as **0000**. The resulting series of codewords, the data codeword sequence, is then processed as described in 7.5 to add error correction codewords to the message. In certain versions of symbol, it may be necessary to add 3, 4 or 7 Remainder Bits (all zeros) to the end of the message, after the final error correction codeword, in order exactly to fill the symbol capacity (see Table 1).

Table 7 — Number of symbol characters and input data capacity for QR Code

Version	Error correction level	Number of data codewords	Number of data bits	Data capacity			
				Numeric	Alphanumeric	Byte	Kanji
M1	Error Detection only	3	20	5	-	-	-
M2	L	5	40	10	6	-	-
	M	4	32	8	5	-	-
M3	L	11	84	23	14	9	6
	M	9	68	18	11	7	4
M4	L	16	128	35	21	15	9
	M	14	112	30	18	13	8
	Q	10	80	21	13	9	5
1	L	19	152	41	25	17	10
	M	16	128	34	20	14	8
	Q	13	104	27	16	11	7
	H	9	72	17	10	7	4
2	L	34	272	77	47	32	20
	M	28	224	63	38	26	16
	Q	22	176	48	29	20	12
	H	16	128	34	20	14	8
3	L	55	440	127	77	53	32
	M	44	352	101	61	42	26
	Q	34	272	77	47	32	20
	H	26	208	58	35	24	15
4	L	80	640	187	114	78	48
	M	64	512	149	90	62	38
	Q	48	384	111	67	46	28
	H	36	288	82	50	34	21
5	L	108	864	255	154	106	65
	M	86	688	202	122	84	52
	Q	62	496	144	87	60	37
	H	46	368	106	64	44	27
6	L	136	1 088	322	195	134	82
	M	108	864	255	154	106	65
	Q	76	608	178	108	74	45
	H	60	480	139	84	58	36
7	L	156	1 248	370	224	154	95
	M	124	992	293	178	122	75
	Q	88	704	207	125	86	53
	H	66	528	154	93	64	39
8	L	194	1 552	461	279	192	118
	M	154	1 232	365	221	152	93
	Q	110	880	259	157	108	66
	H	86	688	202	122	84	52
9	L	232	1 856	552	335	230	141
	M	182	1 456	432	262	180	111
	Q	132	1 056	312	189	130	80
	H	100	800	235	143	98	60

Table 7 *(continued)*

Version	Error correction level	Number of data codewords	Number of data bits	Data capacity			
				Numeric	Alphanumeric	Byte	Kanji
10	L	274	2 192	652	395	271	167
	M	216	1 728	513	311	213	131
	Q	154	1 232	364	221	151	93
	H	122	976	288	174	119	74
11	L	324	2 592	772	468	321	198
	M	254	2 032	604	366	251	155
	Q	180	1 440	427	259	177	109
	H	140	1 120	331	200	137	85
12	L	370	2 960	883	535	367	226
	M	290	2 320	691	419	287	177
	Q	206	1 648	489	296	203	125
	H	158	1 264	374	227	155	96
13	L	428	3 424	1 022	619	425	262
	M	334	2 672	796	483	331	204
	Q	244	1 952	580	352	241	149
	H	180	1 440	427	259	177	109
14	L	461	3 688	1 101	667	458	282
	M	365	2 920	871	528	362	223
	Q	261	2 088	621	376	258	159
	H	197	1 576	468	283	194	120
15	L	523	4 184	1 250	758	520	320
	M	415	3 320	991	600	412	254
	Q	295	2 360	703	426	292	180
	H	223	1 784	530	321	220	136
16	L	589	4 712	1 408	854	586	361
	M	453	3 624	1 082	656	450	277
	Q	325	2 600	775	470	322	198
	H	253	2 024	602	365	250	154
17	L	647	5 176	1 548	938	644	397
	M	507	4 056	1 212	734	504	310
	Q	367	2 936	876	531	364	224
	H	283	2 264	674	408	280	173
18	L	721	5 768	1 725	1 046	718	442
	M	563	4 504	1 346	816	560	345
	Q	397	3 176	948	574	394	243
	H	313	2 504	746	452	310	191
19	L	795	6 360	1 903	1 153	792	488
	M	627	5 016	1 500	909	624	384
	Q	445	3 560	1 063	644	442	272
	H	341	2 728	813	493	338	208
20	L	861	6 888	2 061	1 249	858	528
	M	669	5 352	1 600	970	666	410
	Q	485	3 880	1 159	702	482	297
	H	385	3 080	919	557	382	235
21	L	932	7 456	2 232	1 352	929	572
	M	714	5 712	1 708	1 035	711	438
	Q	512	4 096	1 224	742	509	314
	H	406	3 248	969	587	403	248
22	L	1 006	8 048	2 409	1 460	1 003	618
	M	782	6 256	1 872	1 134	779	480
	Q	568	4 544	1 358	823	565	348
	H	442	3 536	1 056	640	439	270

Table 7 *(continued)*

Version	Error correction level	Number of data codewords	Number of data bits	Data capacity			
				Numeric	Alphanumeric	Byte	Kanji
23	L	1 094	8 752	2 620	1 588	1 091	672
	M	860	6 880	2 059	1 248	857	528
	Q	614	4 912	1 468	890	611	376
	H	464	3 712	1 108	672	461	284
24	L	1 174	9 392	2 812	1 704	1 171	721
	M	914	7 312	2 188	1 326	911	561
	Q	664	5 312	1 588	963	661	407
	H	514	4 112	1 228	744	511	315
25	L	1 276	10 208	3 057	1 853	1 273	784
	M	1 000	8 000	2 395	1 451	997	614
	Q	718	5 744	1 718	1 041	715	440
	H	538	4 304	1 286	779	535	330
26	L	1 370	10 960	3 283	1 990	1 367	842
	M	1 062	8 496	2 544	1 542	1 059	652
	Q	754	6 032	1 804	1 094	751	462
	H	596	4 768	1 425	864	593	365
27	L	1 468	11 744	3 517	2 132	1 465	902
	M	1 128	9 024	2 701	1 637	1 125	692
	Q	808	6 464	1 933	1 172	805	496
	H	628	5 024	1 501	910	625	385
28	L	1 531	12 248	3 669	2 223	1 528	940
	M	1 193	9 544	2 857	1 732	1 190	732
	Q	871	6 968	2 085	1 263	868	534
	H	661	5 288	1 581	958	658	405
29	L	1 631	13 048	3 909	2 369	1 628	1 002
	M	1 267	10 136	3 035	1 839	1 264	778
	Q	911	7 288	2 181	1 322	908	559
	H	701	5 608	1 677	1 016	698	430
30	L	1 735	13 880	4 158	2 520	1 732	1 066
	M	1 373	10 984	3 289	1 994	1 370	843
	Q	985	7 880	2 358	1 429	982	604
	H	745	5 960	1 782	1 080	742	457
31	L	1 843	14 744	4 417	2 677	1 840	1 132
	M	1 455	11 640	3 486	2 113	1 452	894
	Q	1 033	8 264	2 473	1 499	1 030	634
	H	793	6 344	1 897	1 150	790	486
32	L	1 955	15 640	4 686	2 840	1 952	1 201
	M	1 541	12 328	3 693	2 238	1 538	947
	Q	1 115	8 920	2 670	1 618	1 112	684
	H	845	6 760	2 022	1 226	842	518
33	L	2 071	16 568	4 965	3 009	2 068	1 273
	M	1 631	13 048	3 909	2 369	1 628	1 002
	Q	1 171	9 368	2 805	1 700	1 168	719
	H	901	7 208	2 157	1 307	898	553
34	L	2 191	17 528	5 253	3 183	2 188	1 347
	M	1 725	13 800	4 134	2 506	1 722	1 060
	Q	1 231	9 848	2 949	1 787	1 228	756
	H	961	7 688	2 301	1 394	958	590
35	L	2 306	18 448	5 529	3 351	2 303	1 417
	M	1 812	14 496	4 343	2 632	1 809	1 113
	Q	1 286	10 288	3 081	1 867	1 283	790
	H	986	7 888	2 361	1 431	983	605

Table 7 *(continued)*

Version	Error correction level	Number of data codewords	Number of data bits	Data capacity			
				Numeric	Alphanumeric	Byte	Kanji
36	L	2 434	19 472	5 836	3 537	2 431	1 496
	M	1 914	15 312	4 588	2 780	1 911	1 176
	Q	1 354	10 832	3 244	1 966	1 351	832
	H	1 054	8 432	2 524	1 530	1 051	647
37	L	2 566	20 528	6 153	3 729	2 563	1 577
	M	1 992	15 936	4 775	2 894	1 989	1 224
	Q	1 426	11 408	3 417	2 071	1 423	876
	H	1 096	8 768	2 625	1 591	1 093	673
38	L	2 702	21 616	6 479	3 927	2 699	1 661
	M	2 102	16 816	5 039	3 054	2 099	1 292
	Q	1 502	12 016	3 599	2 181	1 499	923
	H	1 142	9 136	2 735	1 658	1 139	701
39	L	2 812	22 496	6 743	4 087	2 809	1 729
	M	2 216	17 728	5 313	3 220	2 213	1 362
	Q	1 582	12 656	3 791	2 298	1 579	972
	H	1 222	9 776	2 927	1 774	1 219	750
40	L	2 956	23 648	7 089	4 296	2 953	1 817
	M	2 334	18 672	5 596	3 391	2 331	1 435
	Q	1 666	13 328	3 993	2 420	1 663	1 024
	H	1 276	10 208	3 057	1 852	1 273	784

NOTE 1 All codewords shall be 8 bits in length, except that the final data codeword for Versions M1 and M3 is 4 bits long.

NOTE 2 The number of Data Bits includes bits for mode indicator and character count indicator.

7.5 Error correction

7.5.1 Error correction capacity

QR Code employs Reed-Solomon error control coding to detect and correct errors. A series of error correction codewords is generated, which are added to the data codeword sequence in order to enable the symbol to withstand damage without loss of data. There are four user-selectable levels of error correction, as shown in Table 8, offering the capability of recovery from the following amounts of damage:

Table 8 — Error correction levels

Error Correction Level	Recovery Capacity % (approx.)
L	7
M	15
Q	25
H	30

Annex K.2 gives guidance on the appropriate level of error correction to be applied to a symbol.

Error correction level H is not available in Micro QR Code symbols.

The error correction codewords can correct two types of erroneous codewords, erasures (erroneous codewords at known locations) and errors (erroneous codewords at unknown locations). An erasure is an unscanned or undecodable symbol character. An error is a misdecoded symbol character. Since QR Code is a matrix symbology, a defect converting a module from dark to light or vice versa will result in the affected symbol character misdecoding as an apparently valid but different codeword. Such an error causing a substitution error in the data requires two error correction codewords to correct it.

The number of erasures and errors correctable is given by the following formula:

$$e + 2t \leq d - p$$

where:

e number of erasures

t number of errors

d number of error correction codewords

p number of misdecode protection codewords

In the general case, $p = 0$. However, if most of the error correction capacity is used to correct erasures, then the possibility of an undetected error is increased. Whenever the number of erasures is more than half the number of error correction codewords, $p = 3$. For small symbols with less than 8 error correction codewords, erasure correction should not be used ($e = 0$ and $p > 0$).

For example, in a version 6-H symbol there is a total of 172 codewords, of which 112 are error correction codewords (leaving 60 data codewords). The 112 error correction codewords can correct 56 misdecodes or substitution errors, i.e. 56/172 or 32,6% of the symbol capacity.

In the formula above, the following values should be assigned to p:

— $p = 3$ in version 1-L and M2-L symbols,

— $p = 2$ in version 1-M, 2-L, M1, M2-M, M3-L, and M4-L symbols,

— $p = 1$ in version 1-Q, 1-H and 3-L symbols,

— $p = 0$ in all other cases.

Where $p > 0$ there are p (i.e. 1, 2 or 3) codewords which act as error detection codewords and prevent transmission of data from symbols where the number of errors exceeds the error correction capacity, e must be less than $d/2$. In a Version 2-L symbol, for example, the total number of codewords is 44; of these, 34 are data codewords and 10 error correction codewords. From Table 9 it can be seen that the error correction capacity is 4 errors (where $e = 0$). Substituting in the formula above,

$$0 + (2 \times 4) = 10 - 2$$

meaning that the correction of the 4 errors requires only 8 error correction codewords; the remaining 2 error correction codewords can therefore detect (but not correct) any additional errors and the symbol would, if there were more than 4 errors, fail to decode.

Depending on the Version and Error Correction Level, the data codeword sequence shall be subdivided into one or more blocks, to each of which the error correction algorithm shall be applied separately. Table 9 lists, for each version and Error Correction Level, the total number of codewords, the total number of error correction codewords, and the structure and number of error correction blocks.

If Remainder Bits are required to fill remaining modules in the symbol capacity for certain symbol versions they shall all be 0 bits.

Table 9 — Error correction characteristics for QR Code

Version	Total number of codewords	Error correction level	Number of error correction codewords	Value of p	Number of error correction blocks	Error correction code per block (c, k, r)[a]
M1	5	Error detection only	2	2	1	(5,3,0)[b]
M2	10	L M	5 6	3 2	1 1	(10,5,1)[b] (10,4,2)[b]
M3	17	L M	6 8	2	1 1	(17,11,2)[b] (17,9,4)
M4	24	L M Q	8 10 14	2 0 0	1 1 1	(24,16,3)[b] (24,14,5) (24,10,7)
1	26	L M Q H	7 10 13 17	3 2 1 1	1 1 1 1	(26,19,2)[b] (26,16,4)[b] (26,13,6)[b] (26,9,8)[b]
2	44	L M Q H	10 16 22 28	2 0 0 0	1 1 1 1	(44,34,4)[b] (44,28,8) (44,22,11) (44,16,14)
3	70	L M Q H	15 26 36 44	1 0 0 0	1 1 2 2	(70,55,7)[b] (70,44,13) (35,17,9) (35,13,11)
4	100	L M Q H	20 36 52 64	0	1 2 2 4	(100,80,10) (50,32,9) (50,24,13) (25,9,8)
5	134	L M Q H	26 48 72 88	0	1 2 2 2 2 2	(134,108,13) (67,43,12) (33,15,9) (34,16,9) (33,11,11) (34,12,11)
6	172	L M Q H	36 64 96 112	0	2 4 4 4	(86,68,9) (43,27,8) (43,19,12) (43,15,14)
7	196	L M Q H	40 72 108 130	0	2 4 2 4 4 1	(98,78,10) (49,31,9) (32,14,9) (33,15,9) (39,13,13) (40,14,13)
8	242	L M Q H	48 88 132 156	0	2 2 2 4 2 4 2	(121,97,12) (60,38,11) (61,39,11) (40,18,11) (41,19,11) (40,14,13) (41,15,13)

[a] c = total number of codewords, k = number of data codewords, r = error correction capacity

[b] Error correction capacity is less than half the number of error correction codewords to reduce the probability of misdecodes

Table 9 *(continued)*

Version	Total number of codewords	Error correction level	Number of error correction codewords	Value of p	Number of error correction blocks	Error correction code per block (c, k, r)[a]
9	292	L M Q H	60 110 160 192	0	2 3 2 4 4 4 4	(146,116,15) (58,36,11) (59,37,11) (36,16,10) (37,17,10) (36,12,12) (37,13,12)
10	346	L M Q H	72 130 192 224	0	2 2 4 1 6 2 6 2	(86,68,9) (87,69,9) (69,43,13) (70,44,13) (43,19,12) (44,20,12) (43,15,14) (44,16,14)
11	404	L M Q H	80 150 224 264	0	4 1 4 4 4 3 8	(101,81,10) (80,50,15) (81,51,15) (50,22,14) (51,23,14) (36,12,12) (37,13,12
12	466	L M Q H	96 176 260 308	0	2 2 6 2 4 6 7 4	(116,92,12) (117,93,12) (58,36,11) (59,37,11) (46,20,13) (47,21,13) (42,14,14) (43,15,14)
13	532	L M Q H	104 198 288 352	0	4 8 1 8 4 12 4	(133,107,13) (59,37,11) (60,38,11) (44,20,12) (45,21,12) (33,11,11) (34,12,11)
14	581	L M Q H	120 216 320 384	0	3 1 4 5 11 5 11 5	(145,115,15) (146,116,15) (64,40,12) (65,41,12) (36,16,10) (37,17,10) (36,12,12) (37,13,12)

[a] c = total number of codewords, k = number of data codewords, r = error correction capacity

[b] Error correction capacity is less than half the number of error correction codewords to reduce the probability of misdecodes

Table 9 *(continued)*

Version	Total number of codewords	Error correction level	Number of error correction codewords	Value of p	Number of error correction blocks	Error correction code per block (c, k, r)[a]
15	655	L M Q H	132 240 360 432	0	5 1 5 5 5 7 11 7	(109,87,11) (110,88,11) (65,41,12) (66,42,12) (54,24,15) (55,25,15) (36,12,12) (37,13,12)
16	733	L M Q H	144 280 408 480	0	5 1 7 3 15 2 3 13	(122,98,12) (123,99,12) (73,45,14) (74,46,14) (43,19,12) (44,20,12) (45,15,15) (46,16,15)
17	815	L M Q H	168 308 448 532	0	1 5 10 1 1 15 2 17	(135,107,14) (136,108,14) (74,46,14) (75,47,14) (50,22,14) (51,23,14) (42,14,14) (43,15,14)
18	901	L M Q H	180 338 504 588	0	5 1 9 4 17 1 2 19	(150,120,15) (151,121,15) (69,43,13) (70,44,13) (50,22,14) (51,23,14) (42,14,14) (43,15,14)
19	991	L M Q H	196 364 546 650	0	3 4 3 11 17 4 9 16	(141,113,14) (142,114,14) (70,44,13) (71,45,13) (47,21,13) (48,22,13) (39,13,13) (40,14,13)
20	1 085	L M Q H	224 416 600 700	0	3 5 3 13 15 5 15 10	(135,107,14) (136,108,14) (67,41,13) (68,42,13) (54,24,15) (55,25,15) (43,15,14) (44,16,14)

[a] c = total number of codewords, k = number of data codewords, r = error correction capacity

[b] Error correction capacity is less than half the number of error correction codewords to reduce the probability of misdecodes

Table 9 *(continued)*

Version	Total number of codewords	Error correction level	Number of error correction codewords	Value of p	Number of error correction blocks	Error correction code per block (c, k, r)[a]
21	1 156	L M Q H	224 442 644 750	0	4 4 17 17 6 19 6	(144,116,14) (145,117,14) (68,42,13) (50,22,14) (51,23,14) (46,16,15) (47,17,15)
22	1 258	L M Q H	252 476 690 816	0	2 7 17 7 16 34	(139,111,14) (140,112,14) (74,46,14) (54,24,15) (55,25,15) (37,13,12)
23	1 364	L M Q H	270 504 750 900	0	4 5 4 14 11 14 16 14	(151,121,15) (152,122,15) (75,47,14) (76,48,14) (54,24,15) (55,25,15) (45,15,15) (46,16,15)
24	1 474	L M Q H	300 560 810 960	0	6 4 6 14 11 16 30 2	(147,117,15) (148,118,15) (73,45,14) (74,46,14) (54,24,15) (55,25,15) (46,16,15) (47,17,15)
25	1 588	L M Q H	312 588 870 1050	0	8 4 8 13 7 22 22 13	(132,106,13) (133,107,13) (75,47,14) (76,48,14) (54,24,15) (55,25,15) (45,15,15) (46,16,15)
26	1 706	L M Q H	336 644 952 1110	0	10 2 19 4 28 6 33 4	(142,114,14) (143,115,14) (74,46,14) (75,47,14) (50,22,14) (51,23,14) (46,16,15) (47,17,15)

[a] c = total number of codewords, k = number of data codewords, r = error correction capacity

[b] Error correction capacity is less than half the number of error correction codewords to reduce the probability of misdecodes

Table 9 *(continued)*

Version	Total number of codewords	Error correction level	Number of error correction codewords	Value of p	Number of error correction blocks	Error correction code per block (c, k, r)[a]
27	1 828	L M Q H	360 700 1 020 1 200	0	8 4 22 3 8 26 12 28	(152,122,15) (153,123,15) (73,45,14) (74,46,14) (53,23,15) (54,24,15) (45,15,15) (46,16,15)
28	1 921	L M Q H	390 728 1 050 1 260	0	3 10 3 23 4 31 11 31	(147,117,15) (148,118,15) (73,45,14) (74,46,14) (54,24,15) (55,25,15) (45,15,15) (46,16,15)
29	2 051	L M Q H	420 784 1 140 1 350	0	7 7 21 7 1 37 19 26	(146,116,15) (147,117,15) (73,45,14) (74,46,14) (53,23,15) (54,24,15) (45,15,15) (46,16,15)
30	2 185	L M Q H	450 812 1 200 1 440	0	5 10 19 10 15 25 23 25	(145,115,15) (146,116,15) (75,47,14) (76,48,14) (54,24,15) (55,25,15) (45,15,15) (46,16,15)
31	2 323	L M Q H	480 868 1 290 1 530	0	13 3 2 29 42 1 23 28	(145,115,15) (146,116,15) (74,46,14) (75,47,14) (54,24,15) (55,25,15) (45,15,15) (46,16,15)
32	2 465	L M Q H	510 924 1 350 1 620	0	17 10 23 10 35 19 35	(145,115,15) (74,46,14) (75,47,14) (54,24,15) (55,25,15) (45,15,15) (46,16,15)

[a] c = total number of codewords, k = number of data codewords, r = error correction capacity

[b] Error correction capacity is less than half the number of error correction codewords to reduce the probability of misdecodes

Table 9 *(continued)*

Version	Total number of codewords	Error correction level	Number of error correction codewords	Value of p	Number of error correction blocks	Error correction code per block (c, k, r)[a]
33	2 611	L M Q H	540 980 1 440 1 710	0	17 1 14 21 29 19 11 46	(145,115,15) (146,116,15) (74,46,14) (75,47,14) (54,24,15) (55,25,15) (45,15,15) (46,16,15)
34	2 761	L M Q H	570 1 036 1 530 1 800	0	13 6 14 23 44 7 59 1	(145,115,15) (146,116,15) (74,46,14) (75,47,14) (54,24,15) (55,25,15) (46,16,15) (47,17,15)
35	2 876	L M Q H	570 1 064 1 590 1 890	0	12 7 12 26 39 14 22 41	(151,121,15) (152,122,15) (75,47,14) (76,48,14) (54,24,15) (55,25,15) (45,15,15) (46,16,15)
36	3 034	L M Q H	600 1 120 1 680 1 980	0	6 14 6 34 46 10 2 64	(151,121,15) (152,122,15) (75,47,14) (76,48,14) (54,24,15) (55,25,15) (45,15,15) (46,16,15)
37	3 196	L M Q H	630 1 204 1 770 2 100	0	17 4 29 14 49 10 24 46	(152,122,15) (153,123,15) (74,46,14) (75,47,14) (54,24,15) (55,25,15) (45,15,15) (46,16,15)
38	3 362	L M Q H	660 1 260 1 860 2 220	0	4 18 13 32 48 14 42 32	(152,122,15) (153,123,15) (74,46,14) (75,47,14) (54,24,15) (55,25,15) (45,15,15) (46,16,15)

[a] c = total number of codewords, k = number of data codewords, r = error correction capacity

[b] Error correction capacity is less than half the number of error correction codewords to reduce the probability of misdecodes

Table 9 *(continued)*

Version	Total number of codewords	Error correction level	Number of error correction codewords	Value of p	Number of error correction blocks	Error correction code per block (c, k, r)[a]
39	3 532	L M Q H	720 1 316 1 950 2 310	0	20 4 40 7 43 22 10 67	(147,117,15) (148,118,15) (75,47,14) (76,48,14) (54,24,15) (55,25,15) (45,15,15) (46,16,15)
40	3 706	L M Q H	750 1 372 2 040 2 430	0	19 6 18 31 34 34 20 61	(148,118,15) (149,119,15) (75,47,14) (76,48,14) (54,24,15) (55,25,15) (45,15,15) (46,16,15)

[a] c = total number of codewords, k = number of data codewords, r = error correction capacity

[b] Error correction capacity is less than half the number of error correction codewords to reduce the probability of misdecodes

7.5.2 Generating the error correction codewords

The data codewords including Pad codewords as necessary shall be divided into the number of blocks shown in Table 9. Error correction codewords shall be calculated for each block and appended to the data codewords.

NOTE Micro QR Code symbols consist of a single block.

The polynomial arithmetic for QR Code shall be calculated using bit-wise modulo 2 arithmetic and byte-wise modulo 100011101 arithmetic. This is a Galois field of 2^8 with 100011101 representing the field's prime modulus polynomial $x^8 + x^4 + x^3 + x^2 +1$.

The data codewords are the coefficients of the terms of a polynomial with the coefficient of the highest term being the first data codeword and that of the lowest power term being the last data codeword before the first error correction codeword.

The error correction codewords are the remainder after dividing the data codewords by a polynomial *g(x)* used for error correction codes (see Annex A). The highest order coefficient of the remainder is the first error correction codeword and the zero power coefficient is the last error correction codeword and the last codeword in the block.

NOTE If this calculation is performed by "long division" the symbol data polynomial must first be multiplied by x^k.

Thirty-six different generator polynomials are used for generating the error correction codewords for QR Code. These are given in Annex A.

This can be implemented by using the division circuit as shown in Figure 14. The registers b_0 through b_{k-1} are initialized as zeros. There are two phases to generate the encoding. In the first phase, with the switch in the down position the data codewords are passed both to the output and the circuit. The first phase is complete after *n* clock pulses. In the second phase ($n + 1 ... n + k$ clock pulses), with the switch in the up position, the error correction codewords $\varepsilon_{k-1} ... \varepsilon_0$ are generated by flushing the registers in order while keeping the data input at 0.

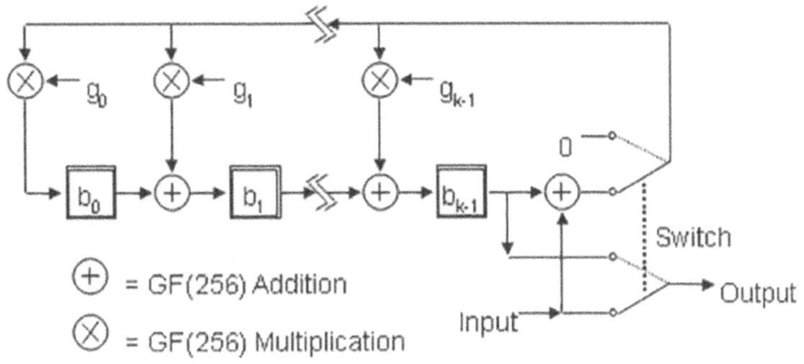

Figure 14 — Error correction codeword encoding circuit

7.6 Constructing the final message codeword sequence

The total number of codewords in the message shall always be equal to the total number of codewords capable of being represented in the symbol, as shown in Tables 7 and 9.

The following steps shall be followed to construct the final sequence of codewords (data plus error correction codewords plus Remainder Codewords if necessary):

1. Divide the data codeword sequence into *n* blocks as defined in Table 9 according to the version and error correction level (or a single block for Micro QR Code symbols).
2. For each data block, calculate a corresponding block of error correction codewords as defined in 7.5.2 and Annex A.
3. Assemble the final sequence by taking data and error correction codewords from each block in turn. For example, if there are four blocks the sequence would be: data block 1, codeword 1; data block 2, codeword 1; ... ; data block 4, codeword 1; data block 1, codeword 2; ... and similarly to data block 3, final codeword; data block 4, final codeword; then error correction block 1, codeword 1, error correction block 2, codeword 1, ... and similarly to error correction block 4, final codeword. QR Code symbols contain data and error correction blocks which always exactly fill the symbol codeword capacity. In certain QR Code versions, however, where the number of modules available for data and error correction codewords is not an exact multiple of 8, there may be a need for 3, 4 or 7 Remainder Bits to be appended to the final message bit stream in order to fill exactly the number of modules in the encoding region.

The shortest data block (or blocks) shall be placed first in the sequence and all the data codewords shall be placed in the symbol before the first error correction codeword. For example, the Version 5-H symbol comprises four data and four error correction blocks, the first two of each of which contain 11 data and 22 error correction codewords respectively, while the third and fourth pairs of blocks contain 12 data and 22 error correction codewords respectively. In this symbol, the character arrangement can be depicted as shown in Figure 15. Each row of the figure corresponds to one block of data codewords (shown as D_n) followed by the associated block of error correction codewords (shown as E_n); the sequence of character placement in the symbol is obtained by reading down each column of the figure in turn.

	Data codewords					Error correction codewords			
Block 1	D_1	D_2	.....	D_{11}		E_1	E_2	.....	E_{22}
Block 2	D_{12}	D_{13}	.....	D_{22}		E_{23}	E_{24}	.....	E_{44}
Block 3	D_{23}	D_{24}	.....	D_{33}	D_{34}	E_{45}	E_{46}	.....	E_{66}
Block 4	D_{35}	D_{36}	.....	D_{45}	D_{46}	E_{67}	E_{68}	.....	E_{88}

Figure 15 — Constructing the final message codeword sequence

The final message codeword sequence for the Version 5-H symbol is therefore:

D_1, D_{12}, D_{23}, D_{35}, D_2, D_{13}, D_{24}, D_{36}, ... D_{11}, D_{22}, D_{33}, D_{45}, D_{34}, D_{46}, E_1, E_{23}, E_{45}, E_{67}, E_2, E_{24}, E_{46}, E_{68}, ... E_{22}, E_{44}, E_{66}, E_{88}. The symbol module capacity is filled by adding 7 Remainder (**0**) bits as needed after the final codeword.

7.7 Codeword placement in matrix

7.7.1 Symbol character representation

There are two types of symbol character, regular and irregular, in the QR Code symbol. Their use depends on their position in the symbol, relative to other symbol characters and function patterns.

Most codewords shall be represented in a regular 2 × 4 module block in the symbol. There are two ways of positioning these blocks, in a vertical arrangement (2 modules wide and 4 modules high) and, if necessary when placement changes direction, in a horizontal arrangement (4 modules wide and 2 modules high). Irregular symbol characters are used when changing direction or in the vicinity of alignment or other function Patterns. Examples are shown in Figures 16, 17 and 18.

7.7.2 Function pattern placement

A square blank matrix shall be constructed with the number of modules horizontally and vertically corresponding to the Version in use. Positions corresponding to the finder pattern, separator, timing pattern, and alignment patterns shall be filled with either dark modules or light modules as appropriate. Module positions for the format information and version information shall be left temporarily blank. These blank positions are shown in Figures 19 and 20 and are common to all Versions (although the version information is not present in Version 1 to 6 symbols). Annex E defines the positioning of alignment patterns.

7.7.3 Symbol character placement

In the encoding region of the QR Code symbol, symbol characters are positioned in two-module wide columns commencing at the lower right corner of the symbol and running alternately upwards and downwards from the right to the left. The principles governing the placement of characters and of bits within the characters are given below. Figures 19 and 20 illustrate Version 2 and Version 7 symbols applying these principles.

a) The sequence of bit placement in the column shall be from right to left and either upwards or downwards in accordance with the direction of symbol character placement.

b) The most significant bit (shown as bit 7) of each codeword shall be placed in the first available module position. Subsequent bits shall be placed in the next module positions. The most significant bit therefore occupies the lower right module of a regular symbol character when the direction of placement is upwards, and the upper right module when the direction of placement is downwards. It may however occupy the lower left module of an irregular symbol character if the previous character has ended in the right-hand module column (see Figure 18).

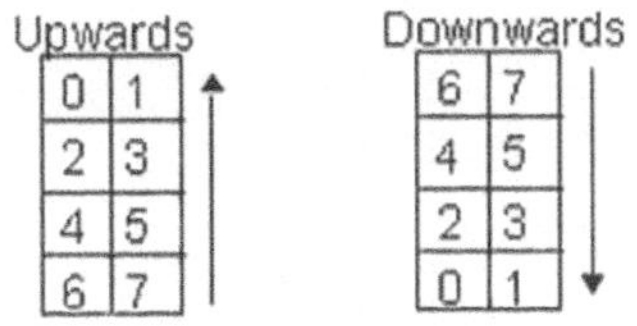

Figure 16 — Bit placement in regular symbol character in upwards and downwards directions

c) When a symbol character encounters the horizontal boundary of an alignment pattern or of the timing pattern in both module columns, it shall continue above or below the pattern as though the encoding region were continuous.

d) When the upper or lower boundary of the symbol character region is reached (i.e. the edge of the symbol, format information, version information, or separator) any remaining bits in the codeword shall be placed in the next column to the left. The direction of placement reverses.

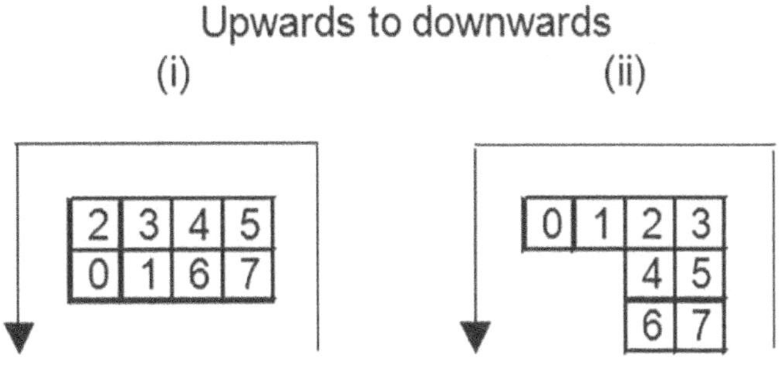

Figure 17 — Example of bit placement in (i) regular and (ii) irregular symbol characters when direction of placement changes

e) When the right-hand module column of the symbol character column encounters an alignment pattern or an area occupied by version information, bits are placed to form an irregular symbol character, extending along the single module column adjacent to the alignment pattern or version information. If the character ends before two columns are available for the next symbol character, the most significant bit of the next character shall be placed in the single column.

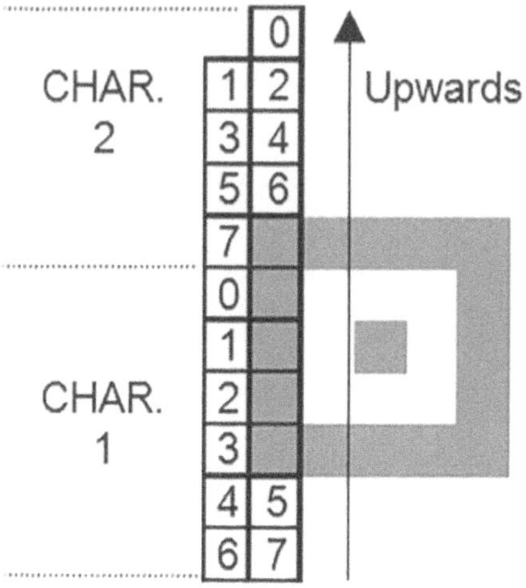

Figure 18 — Example of bit placement adjacent to alignment pattern

An alternative method for placement in the symbol, which yields the same result, is to regard the interleaved codeword sequence as a single bit stream, which is placed (starting with the most significant bit) in the two-module wide columns alternately upwards and downwards from the right to left of the symbol. In each column the bits are placed alternately in the right and left modules, moving upwards or downwards according to the direction of placement and skipping areas occupied by function patterns, changing direction at the top or bottom of the column. Each bit shall always be placed in the first available module position.

When the data capacity of the symbol is such that it does not divide exactly into a number of 8-bit symbol characters, the appropriate number of Remainder Bits (3, 4 or 7 as shown in Table 1) shall be used to fill the symbol capacity. These Remainder Bits shall always have the value 0 before data masking according to 7.8.

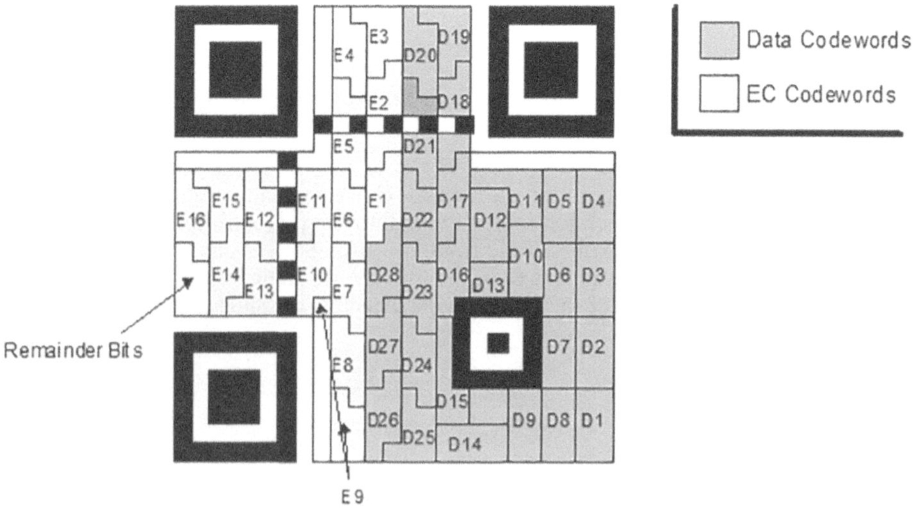

Figure 19 — Symbol character arrangement in version 2-M symbol

D1 - D13	Data Block 1
D14 - D26	Data Block 2
D27 - D39	Data Block 3
D40 - D52	Data Block 4
D53 - D66	Data Block 5
E1 - E26	EC Block 1
E27 - E52	EC Block 2
E53 - E78	EC Block 3
E79 - E104	EC Block 4
E105 - E130	EC Block 5

Figure 20 — Symbol character arrangement in version 7-H symbol

Exactly the same principles apply to a Micro QR Code symbol. There are no irregular symbol characters in these symbols and the sole exception is that D_3 in a Version M1 symbol, D_{11} in a Version M3-L symbol and D_9 in a Version M3-M symbol is a 2 × 2 square 4-module block.

7.8 Data masking

7.8.1 General

For reliable QR Code reading, it is preferable for dark and light modules to be arranged in a well-balanced manner in the symbol. The module pattern **1011101** particularly found in the finder pattern should be avoided in other areas of the symbol as much as possible. To meet the above conditions, data masking should be applied following the steps described below:

1. Data masking is not applied to function patterns.
2. Convert the given module pattern in the encoding region (excluding the format information and the version information) with multiple matrix patterns successively through the XOR operation. For the XOR operation, lay the module pattern over each of the data masking matrix patterns in turn and reverse the modules (from light to dark or vice versa) which correspond to dark modules of the data masking pattern.
3. Then evaluate all the resulting converted patterns by charging penalties for undesirable features on each conversion result.
4. Select the pattern with the lowest penalty points score.

7.8.2 Data mask patterns

Table 10 shows the data mask pattern reference (binary reference for use in the format information) and the data mask pattern generation condition. The data mask pattern is generated by defining as dark any module in the encoding region (excluding the area reserved for format information and the version information) for which the condition is true; in the condition, *i* refers to the row position of the module in question and *j* to its column position, with $(i, j) = (0, 0)$ for the top left module in the symbol.

Table 10 — Data mask pattern generation conditions

Data mask pattern reference for QR Code symbols	Data mask pattern reference for Micro QR Code symbols	Condition
000		$(i + j) \bmod 2 = 0$
001	00	$i \bmod 2 = 0$
010		$j \bmod 3 = 0$
011		$(i + j) \bmod 3 = 0$
100	01	$((i \operatorname{div} 2) + (j \operatorname{div} 3)) \bmod 2 = 0$
101		$(i j) \bmod 2 + (i j) \bmod 3 = 0$
110	10	$((i j) \bmod 2 + (i j) \bmod 3) \bmod 2 = 0$
111	11	$((i+j) \bmod 2 + (i j) \bmod 3) \bmod 2 = 0$

Figure 21 shows all data mask patterns, illustrated in a version 1 symbol. Figure 23 simulates the effects of data masking using data mask pattern references 000 to 111.

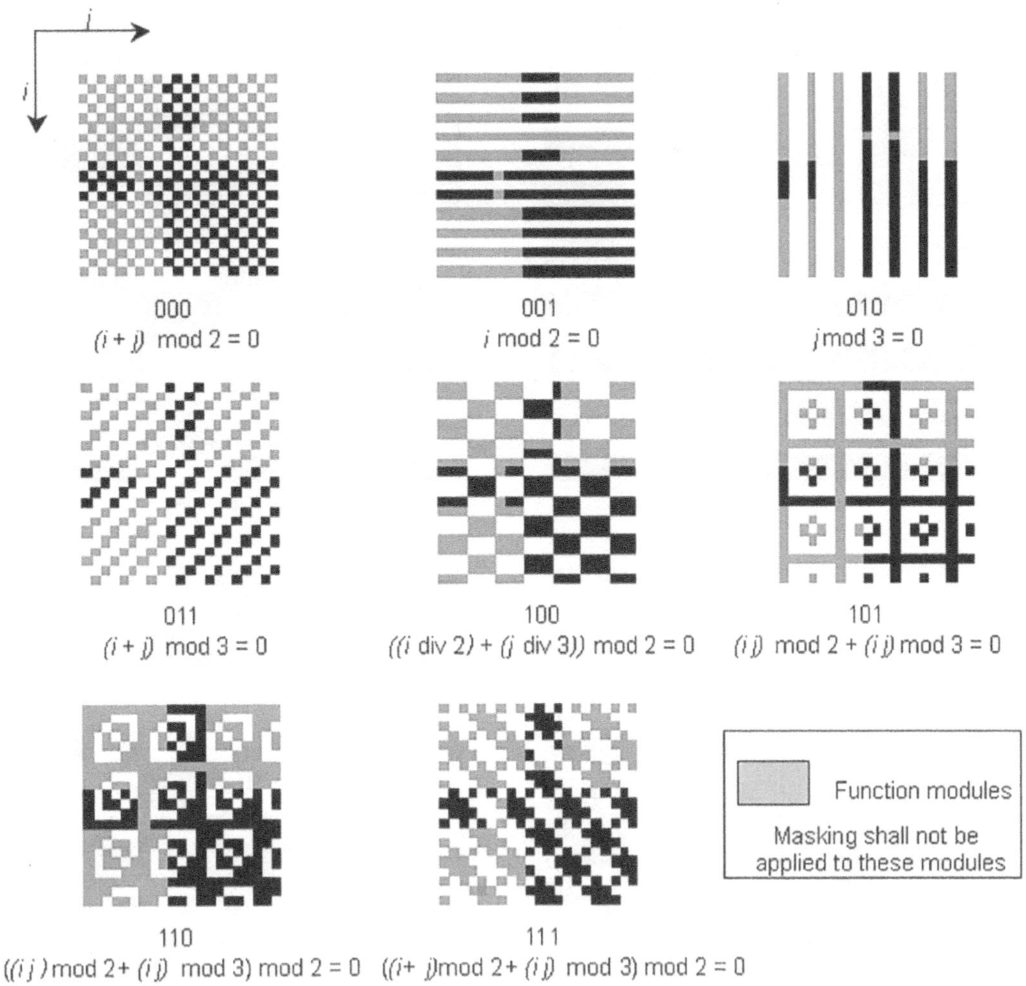

Figure 21 — Data mask patterns for version 1 symbol

NOTE 1 The three bits below each pattern represent the data mask pattern reference.

NOTE 2 The equation below the data mask pattern reference shows the data mask pattern generation condition; modules which meet the condition are shown dark.

Figure 22 below shows the four available data masking patterns applied to a Micro QR Code version M-4 symbol.

Figure 22 — Data mask patterns applied to Micro QR Code version M4 symbol

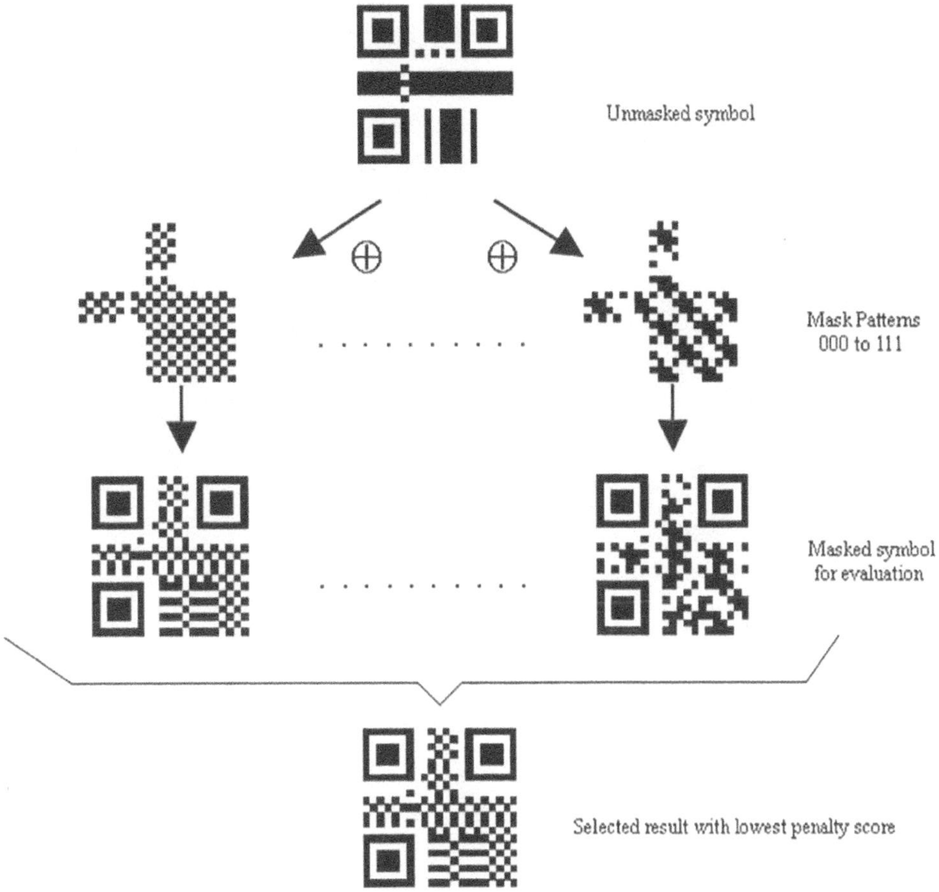

Figure 23 — Data masking simulation in QR Code symbols

7.8.3 Evaluation of data masking results

7.8.3.1 Evaluation of QR Code symbols

After performing the data masking operation with each data mask pattern in turn, the results shall be evaluated by scoring penalty points for each occurrence of the following features. The higher the number of points, the less acceptable the result. In Table 11 below, the variables N_1 to N_4 represent weighted penalty scores for the undesirable features ($N_1 = 3$, $N_2 = 3$, $N_3 = 40$, $N_4 = 10$), i is the amount by which the number of adjacent modules of the same color exceeds 5 and k is the rating of the deviation of the proportion of dark modules in the symbol from 50 % in steps of 5 %. Although the data masking operation is only performed on the encoding region of the symbol excluding the format information, the area to be evaluated is the complete symbol.

Table 11 — Scoring of data masking results

Feature	Evaluation condition	Points
Adjacent modules in row/column in same color	No. of modules = (5 + i)	$N_1 + i$
Block of modules in same color	Block size = 2 × 2	N_2
1 : 1 : 3 : 1 : 1 ratio (dark:light:dark:light:dark) pattern in row/column, preceded or followed by light area 4 modules wide	Existence of the pattern	N_3
Proportion of dark modules in entire symbol	50 × (5 × k)% to 50 × (5 × (k + 1))%	$N_4 \times k$

NOTE 1 Adjacent modules in row/column in the same colour.

Check the blocks consisting of light (white) or dark (black) modules of more than five in a row both laterally and vertically for the evaluation of data masking results. The rule of this calculation is that 3 penalty points shall be added to each block of five consecutive modules, 4 penalty points for each block of six consecutive modules and so on, with scoring by 1 point each time the number of modules increases. For example, impose 5 penalty points on the block of "dark:dark:dark:dark:dark:dar k:dark" module pattern, where a series of seven consecutive modules is counted as one block. However, do not double-count the point. The penalty point for a seven-module block, for example, shall be 5, not the sum of 3 (for a five-module block) + 4 (for a six-module block) + 5 (for a seven-module block) =12.

NOTE 2 Module blocks in the same colour.

The penalty point shall be equal to the number of blocks with 2 x 2 light or dark modules. Take a block consisting of 3 x 3 dark modules for an example. Considering that up to four 2 x 2 dark modules can be included in this block, the penalty applied to this block shall be calculated as 4 (blocks) x 3 (points) = 12 points.

NOTE 3 1:1:3:1:1 ratio pattern in row/column.

If the light area of more than 4 module wide exists after or before a 1:1:3:1:1 ratio (dark:light:dark:light:dark) pattern, the imposed penalty shall be 40 points.

NOTE 4 Proportion of dark modules in entire symbol.

Add 10 points to a deviation of 5% increment or decrement in the proportion ratio of dark module from the referential 50% (or 0 point) level. For example, assign 0 points as a penalty if the ratio of dark module is between 45% and 55%, or 10 points if the ratio of dark module is between 40% and 60%.

The data mask pattern which results in the lowest penalty score shall be selected for the symbol.

7.8.3.2 Evaluation of Micro QR Code symbols

After performing the data masking operation on the encoding region of the symbol with each data mask pattern in turn, the results shall be evaluated by scoring points for the number of dark modules in each of the two edges which are not timing patterns. The lower the number of points, the less acceptable the result. In these symbols, it is desirable to have more dark modules in the edge, in order to differentiate a quiet zone from an encoding region more effectively.

For each data mask pattern in turn, count the number of dark modules in the right and lower edges of the symbol (excluding the final module of the timing pattern). The evaluation score is given by the following formula:

If $SUM_1 \leq SUM_2$

$$\text{Evaluation score} = SUM_1 \times 16 + SUM_2$$

If $SUM_1 > SUM_2$

$$\text{Evaluation score} = SUM_2 \times 16 + SUM_1$$

where:

SUM_1 number of dark modules in right side edge

SUM_2 number of dark modules in lower side edge

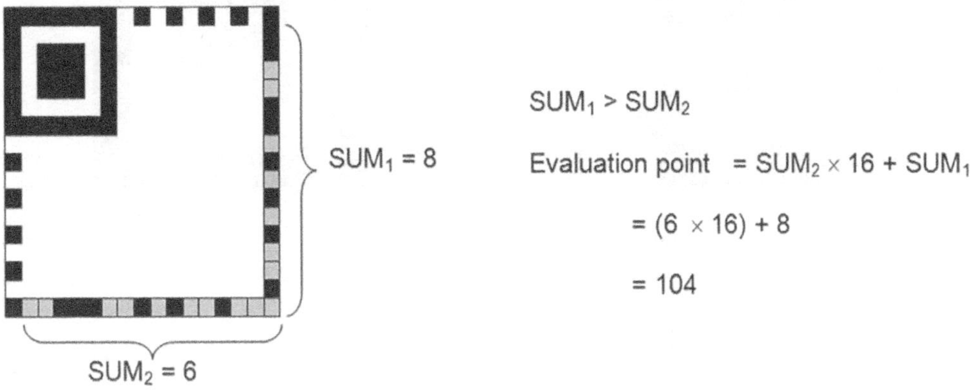

Figure 24 — Evaluation of masking results in Micro QR Code symbol

The data mask pattern which results in the highest score shall be selected for the symbol.

7.9 Format information

7.9.1 QR Code symbols

The format information is a 15-bit sequence containing 5 data bits, with 10 error correction bits calculated using the (15, 5) BCH code. For details of the error correction calculation for the format information, refer to Annex C. The first two data bits contain the Error Correction Level of the symbol, indicated in Table 12.

Table 12 — Error correction level indicators for QR Code symbols

Error Correction Level	Binary indicator
L	01
M	00
Q	11
H	10

The third to fifth data bits of the format information contain the data mask pattern reference from Table 10 above for the pattern selected according to 7.8.3.

The 10 error correction bits shall be calculated as described in Annex C and appended to the 5 data bits.

The 15-bit error corrected format information shall then be XORed with the Mask Pattern **101010000010010**, in order to ensure that no combination of Error Correction Level and data mask pattern will result in an all-zero data string.

The resulting masked format information shall be mapped into the areas reserved for it in the symbol as shown in Figure 25. Note that the format information appears twice in the symbol in order to provide redundancy since its correct decoding is essential to the decoding of the complete symbol. The least significant bit of the format information is located in the modules numbered 0, and the most significant bit in the modules numbered 14 in Figure 25. The module in position (4V + 9, 8) where V is the version number, shall always be dark and does not form part of the format information.

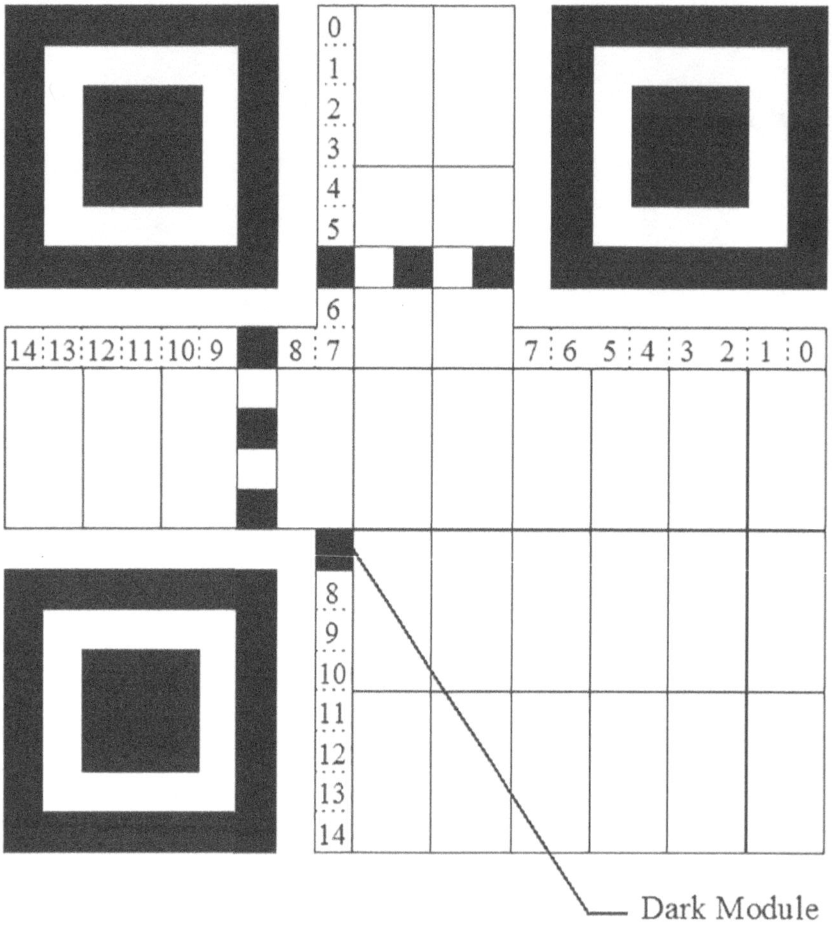

EXAMPLE

Assume Error Correction Level M	**00**
and data mask pattern reference:	**101**
Data:	**00101**
BCH bits:	**0011011100**
Unmasked bit sequence:	**001010011011100**
Mask pattern for XOR operation:	**101010000010010**
Format information module pattern:	**100000011001110**

bit 14 bit 0

Figure 25 — Format information positioning

7.9.2 Micro QR Code symbols

The format information is a 15-bit sequence containing 5 data bits, with 10 error correction bits calculated using the (15, 5) BCH code. For details of the error correction calculation for the format information, refer to Annex C. The first three data bits contain the symbol number (in binary), which identifies the version and error correction level, as shown in Table 13:

Table 13 — Symbol numbers for Micro QR Code symbols

Symbol number	Version	Error Correction Level	Binary Indicator
0	M1	Error detection only	000
1	M2	L	001
2	M2	M	010
3	M3	L	011
4	M3	M	100
5	M4	L	101
6	M4	M	110
7	M4	Q	111

The fourth and fifth data bits of the format information contain the data mask pattern reference shown in Table 10 for the pattern selected according to 7.8.3.

The 10 error correction bits shall be calculated as described in Annex C and appended to the 5 data bits.

The 15-bit error corrected format information shall then be XORed with the bit pattern **100010001000101**, in order to ensure that no combination of symbol number and data mask pattern will result in an all-zero data string.

The resulting masked format information shall be mapped into the areas reserved for it in the symbol as shown in Figure 25 or 26, depending on the symbol type. The least significant bit of the format information is located in the module numbered 0, and the most significant bit in the module numbered 14 in Figures 24 and 25.

EXAMPLE

Symbol number 0: **000**

Data mask pattern reference:	**11**
Data bits (symbol number, data mask pattern reference):	**00011**
BCH bits:	**1101011001**
Unmasked bit sequence:	**000111101011001**
Mask pattern for XOR operation:	**100010001000101**
Format information module pattern:	**100101100011100**

bit 14 … bit 0

Figure 26 — Micro QR Code symbol format information bit positions

7.10 Version information

The version information is included in QR Code symbols of version 7 or larger. It consists of an 18-bit sequence containing 6 data bits, with 12 error correction bits calculated using the (18, 6) Golay code. For details of the error correction calculation for the version information, refer to Annex D. The six data bits contain the Version of the symbol, most significant bit first.

The 12 error correction bits shall be calculated as described in Annex D and appended to the 6 data bits.

No version information will result in an all-zero data string since only Versions 7 to 40 symbols contain the version information. Masking is not therefore applied to the version information.

The resulting version information shall be mapped into the areas reserved for it in the symbol as shown in Figure 27. Note that the version information appears twice in the symbol in order to provide redundancy since its correct decoding is essential to the decoding of the complete symbol. The least significant bit of the version information is located in the modules numbered 0, and the most significant bit in the modules numbered 17, in Figure 28.

<u>Example</u>:

Version number:	**7**
Data:	**000111**
BCH bits:	**110010010100**
Version information module pattern:	**000111110010010100**

The version information areas are the 6 × 3 module block above the timing pattern and immediately to the left of the top right finder pattern separator, and the 3 × 6 module block to the left of the timing pattern and immediately above the lower left finder pattern separator.

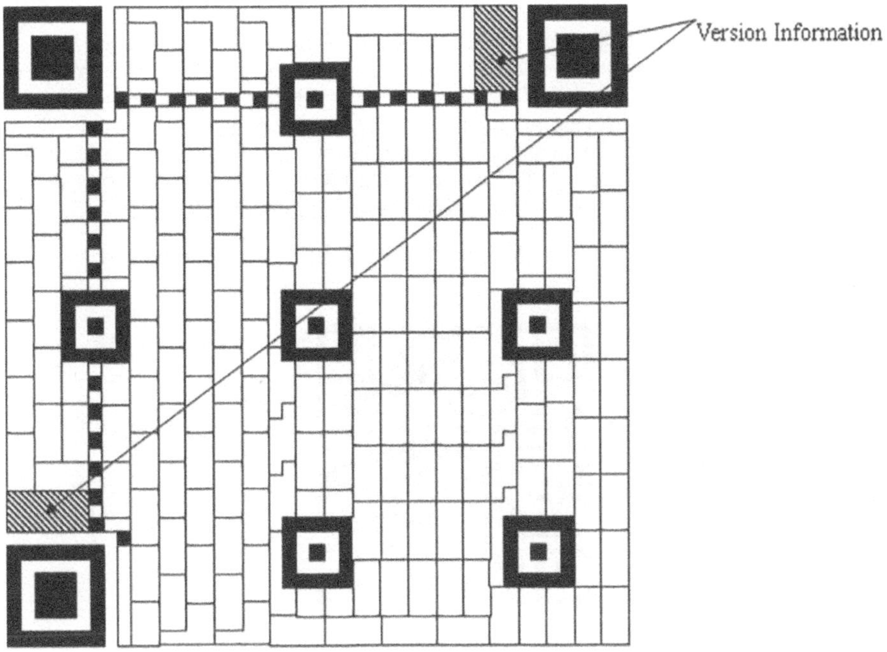

Figure 27 — Version information positioning

0	3	6	9	12	15
1	4	7	10	13	16
2	5	8	11	14	17

Version Information in lower left

0	1	2
3	4	5
6	7	8
9	10	11
12	13	14
15	16	17

Version Information in upper right

Figure 28 — Module arrangement in version information

8 Structured Append

8.1 Basic principles

Structured Append is not available with Micro QR Code symbols.

Up to 16 QR Code symbols may be appended in a structured format. If a symbol is part of a Structured Append message, it is indicated by a header block in the first two and a half symbol character positions.

The Structured Append mode indicator **0011** is placed in the four most significant bit positions in the first symbol character.

This is immediately followed by two Structured Append codewords, spread over the four least significant bits of the first symbol character, the second symbol character and the four most significant bits of the third symbol character. The first codeword is the symbol sequence indicator (see 7.2). The second codeword is the parity data (see 7.3) and is identical in all symbols in the message, enabling it to be verified that all symbols read form part of the same Structured Append message. This header is immediately followed by the data codewords for the symbol commencing with the first mode indicator. If one or more ECIs other than the default ECI is in force, an ECI header for each ECI, consisting of the ECI mode indicator and ECI Designator, shall follow the Structured Append header.

The lower part of Figure 29 shows an example of four Structured Append symbols, with the same data as that in the upper symbol.

Figure 29 — Single symbol (above) and Structured Append series of symbols (below) encoding "ABCDEFGHIJKLMNOPQRSTUVWXYZ0123456789ABCDEFGHIJKLMNOPQRSTUVWXYZ"

8.2 Symbol Sequence Indicator

This codeword indicates the position of the symbol within the set of (up to 16) QR Code symbols in the Structured Append format (in the form *m* of *n* symbols). The first 4 bits of this codeword identify the position of the particular symbol. The last 4 bits identify the total number of symbols to be concatenated in the Structured Append format. The 4-bit patterns shall be the binary equivalents of (m - 1) and (n - 1) respectively.

EXAMPLE

To indicate the 3rd symbol of a set of 7, this shall be encoded thus:

3rd position:	**0010**
Total 7 symbols:	**0110**
Bit pattern:	**00100110**

8.3 Parity Data

The Parity Data shall be an 8-bit byte following the Symbol Sequence Indicator. The parity data is a value obtained by XORing byte by byte the byte values of all the original input data before division into symbol blocks. Mode Indicators, Character Count Identifiers, padding bits, Terminator and Pad Characters shall be excluded from the calculation. Input data is represented for this calculation by 2-byte Shift JIS values for Kanji (each byte being treated separately in the XOR calculation, most significant first) and 8-bit values as shown in Table 6 for other characters. In ECI mode the byte values obtained after any encryption or compression of the data shall be used for the calculation.

For example, "0123456789日本" is divided into "0123", "4567" and "89日本" as follows:

1st symbol block ("0123") - hex. values 30, 31, 32, 33

2nd symbol block ("4567") - hex. values 34, 35, 36, 37

3rd symbol block ("89日本") - hex. values 38, 39, 93FA, 967B

The parity data is calculated from "0123456789日本" by XORing the data successively, byte by byte.

$$30 \oplus 31 \oplus 32 \oplus 33 \oplus 34 \oplus 35 \oplus 36 \oplus 37 \oplus 38 \oplus 39 \oplus 93 \oplus \mathbf{FA} \oplus 96 \oplus \mathbf{7B} = \mathbf{85}$$

Note that the calculation of the parity data may be performed either before the data is sent to the printer or in the printer, based on the capabilities of the printer.

9 Symbol printing and marking

9.1 Dimensions

QR Code symbols shall conform to the following dimensions:

X dimension: the width of a module shall be specified by the application, taking into account the scanning technology to be used, and the technology to produce the symbol;

Y dimension: the height of a module shall be equal to the X dimension;

minimum quiet zone: equal to 2X (for Micro QR Code symbols) or 4X (for QR Code symbols) wide on all four sides.

9.2 Human-readable interpretation

Because QR Code symbols are capable of encoding thousands of characters, a human readable interpretation of the data characters may not be practical. As an alternative, descriptive text rather than literal text may accompany the symbol.

The character size and font are not specified, and the message may be printed anywhere in the area surrounding the symbol. The human readable interpretation should not interfere with the symbol itself nor the quiet zones.

9.3 Marking guidelines

QR Code symbols can be printed or marked using a number of different techniques. Annex K provides user guidelines.

10 Symbol quality

10.1 Methodology

QR Code symbols shall be assessed for quality using the 2D matrix bar code symbol print quality guidelines defined in ISO/IEC 15415, as augmented and modified below.

Some marking technologies may not be able to produce symbols conforming to this International Standard without taking special precautions. Annex M gives additional guidance to help any printing system achieve valid QR Code symbols.

Directly marked symbols (DPM) and/or symbols printed with disconnected dots may not pass this methodology and may not be readable by QR Code scanners. Application requiring such symbols should specify quality measurement using the ISO/IEC 15415 quality extension ISO/IEC/TR 29158 and may require specialized DPM scanners.

10.2 Symbol quality parameters

10.2.1 Fixed pattern damage

Annex G defines the measurement and grading basis for Fixed Pattern Damage.

10.2.2 Scan grade and overall symbol grade

The scan grade shall be the lowest of the grades for symbol contrast, modulation, fixed pattern damage, decode, axial non-uniformity, grid non-uniformity and unused error correction in an individual image of the symbol. The overall symbol grade is the arithmetic mean of the individual scan grades for a number of tested images of the symbol.

10.2.3 Grid non-uniformity

The ideal grid is calculated by using the finder patterns and alignment patterns as datum points, as located by the use of the reference decode algorithm (see Clause 12).

10.3 Process control measurements

A variety of tools and methods can be used to perform useful measurements for monitoring and controlling the process of creating QR Code symbols. These are described in Annex M. These techniques do not constitute a print quality check of the produced symbols (the method specified earlier in this clause and Annex G is the required method for assessing symbol print quality) but they individually and collectively yield good indications of whether the symbol print process is creating workable symbols.

11 Decoding procedure overview

The decoding steps from reading a QR Code symbol to outputting data characters are the reverse of the encoding procedure. Figure 30 shows an outline of the process flow.

1. Locate and obtain an image of the symbol. Recognize dark and light modules as an array of "0" and "1" bits. Identify reflectance polarity from finder pattern module colouring.

2. Read the format information. Release the masking pattern and perform error correction on the format information modules as necessary; if successful, symbol is in normal orientation, otherwise attempt mirror image decoding of format information. Identify Error Correction Level, either directly, in QR Code symbols, or from Micro QR Code symbol number, and data mask pattern reference.

3. Read the version information (where applicable), then determine the version of the symbol (from symbol number, in the case of Micro QR Code symbols).

4. Release the data masking by XORing the encoding region bit pattern with the data mask pattern the reference of which has been extracted from the format information.
5. Read the symbol characters according to the placement rules for the model and restore the data and error correction codewords of the message.
6. Detect errors using the error correction codewords corresponding to the Level Information. If any error is detected, correct it.
7. Divide the data codewords into segments according to the mode indicators and character count indicators.
8. Finally, decode the Data Characters in accordance with the mode(s) in use and output the result.

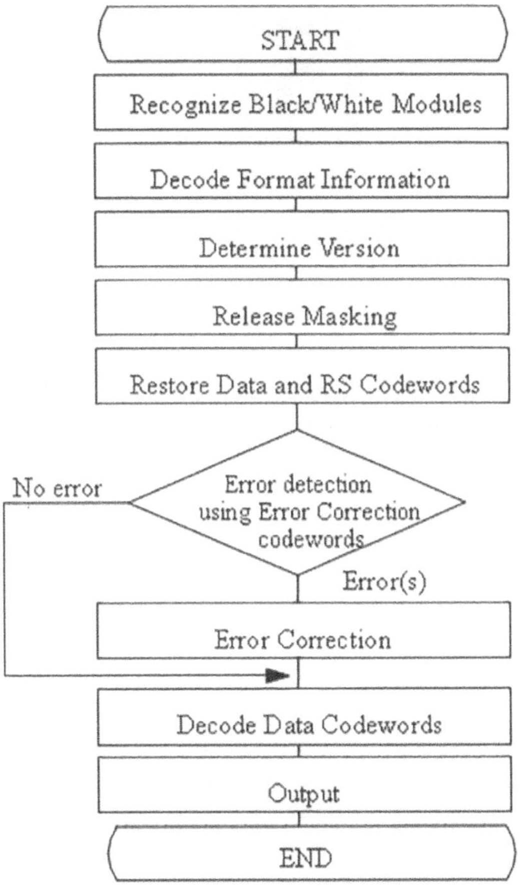

Figure 30 — QR Code decoding steps

12 Reference decode algorithm for QR Code

This reference decode algorithm finds the symbol in an image and decodes it. The decode algorithm refers to dark and light states in the image.

a) Determine a Global Threshold by taking a reflectance value midway between the maximum reflectance and minimum reflectance in the image. Convert the image to a set of dark and light pixels using the Global Threshold.

b) Locate the finder pattern. The finder pattern in QR Code consists of three identical finder patterns located at three of the four corners of the symbol. The finder pattern in Micro QR Code is a single finder pattern. As described in 6.3.3, module widths in each finder pattern form a dark-light-dark-light-dark sequence the relative widths of each element of which are in the ratios 1 : 1 : 3 : 1 : 1. For the purposes of this algorithm the tolerance for each of these widths is 0,5 (i.e. a range of 0,5 to 1,5 for the single module box and 2,5 to 3,5 for the three module square box).

   1) When a candidate area is detected note the position of the first and last points A and B respectively at which a line of pixels in the image encounters the outer edges of the finder pattern (see Figure 31). Repeat this for adjacent pixel lines in the image until all lines crossing the central box of the finder pattern in the x axis of the image have been identified.

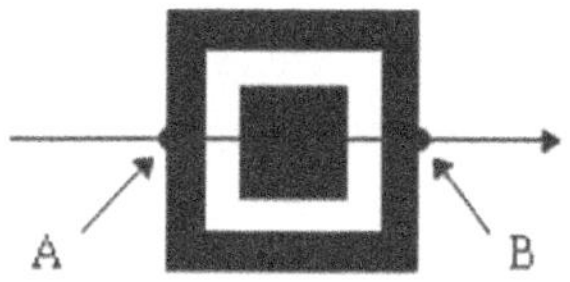

Figure 31 — Scan line in finder pattern

   2) Repeat step 1) for pixel columns crossing the central box of the finder pattern in the y axis of the image.

   3) Locate the center of the pattern. Construct a line through the midpoints between the points A and B on the outermost pixel lines crossing the central box of the finder pattern in the x axis. Construct a similar line through points A and B on the outermost pixel columns crossing the central box in the y axis. The center of the pattern is located at the intersection of these two lines.

   4) Repeat steps 1) to 3) to locate the centers of the two other finder patterns.

   5) If no candidate areas are detected, reverse the colouring of the light and dark pixels and recommence at the beginning of step b to attempt to decode the symbol as a symbol with reflectance reversal.

   6) If a single pattern is identified but two further finder patterns cannot be located, attempt to decode the symbol as a Micro QR Code symbol by jumping to the Micro QR Code symbols reference decode (from step m).

c) Determine the rotational orientation of the symbol by analysing the finder pattern center coordinates to identify which pattern is the upper left pattern in the symbol and the angle of rotation of the symbol.

d) Determine 1) the distance D crossing the full width of the symbol between the centers of the upper left finder pattern and of the upper right finder pattern and 2) the width of the two patterns, W_{UL} and W_{UR} as shown in Figure 32.

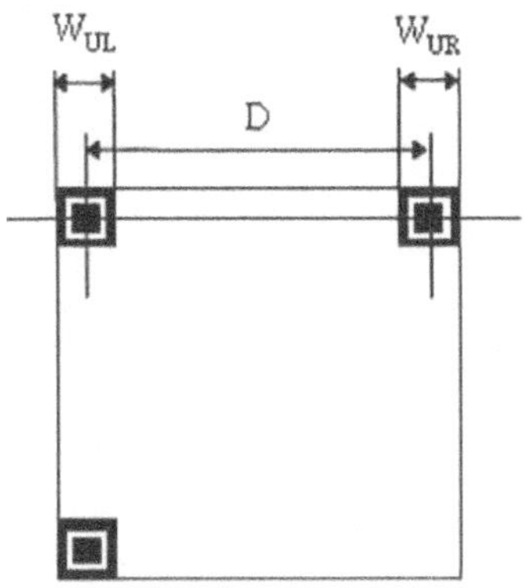

Figure 32 — Upper finder patterns

e) Calculate the nominal X dimension of the symbol.

$$X = (W_{UL} + W_{UR}) / 14$$

f) Provisionally determine the version V of the symbol.

$$V = [(D / X) - 10] / 4$$

g) If the provisional symbol version is 6 or less, this is specified as the defined version. If the provisional symbol version is 7 or more, the version information is decoded as follows.

1) Divide the width W_{UR} of the upper right finder pattern by 7 to calculate the module size CP_{UR}.

$$CP_{UR} = W_{UR} / 7$$

2) Find the guide lines AC and AB from A, B and C, which pass through the centers of the three finder patterns, as shown in Figure 33 below. The sampling grid for each module center in the version information 1 area is determined based on lines parallel to the guide lines, the central coordinates of the finder patterns, and the module size CP_{UR}. Binary values 0 and 1 are determined from the light or dark pattern on the sampling grid.

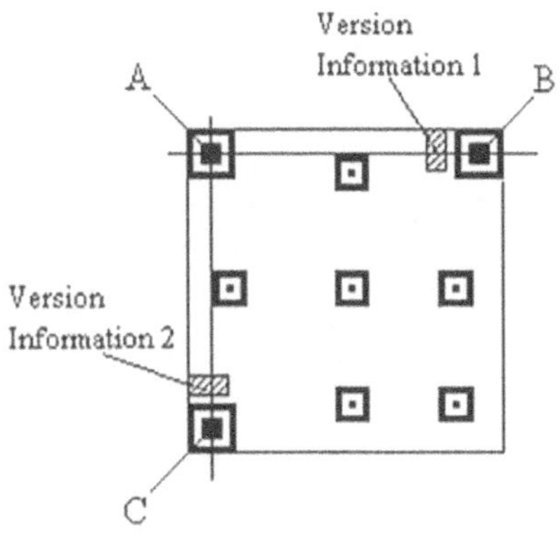

Figure 33 — Finder patterns and version information

3) Determine the version by detecting and correcting errors, if any, based on the table in Annex D.2.

4) If errors exceeding the error correction capacity are detected, then calculate the pattern width W_{DL} of the lower left finder pattern and follow a similar procedure to steps a), b) and c) above to decode version information 2.

h) For Version 1 symbols, redefine X as the average spacing of the center points of the dark and light modules in the upper side Timing Patterns. In a similar manner, calculate the Y dimension as the average spacing of the center points of the dark and light modules in the left side Timing Pattern. Establish a sampling grid based on 1) the horizontal line through the upper side Timing Pattern with lines parallel to it at the vertical spacing of Y, comprising six lines above the horizontal reference line and as many lines below it as are required for the version of the symbol and 2) the vertical line passing through the left side Timing Pattern with lines parallel to it at the horizontal spacing of X, comprising six lines to the left of the vertical reference line and as many lines to the right of it as are required for the version of the symbol. For version 2 and larger symbols, determine the central coordinate of each alignment pattern from the coordinates defined in 6.3.6 and Annex E and construct the sampling grids with lines equidistantly spaced between these points.

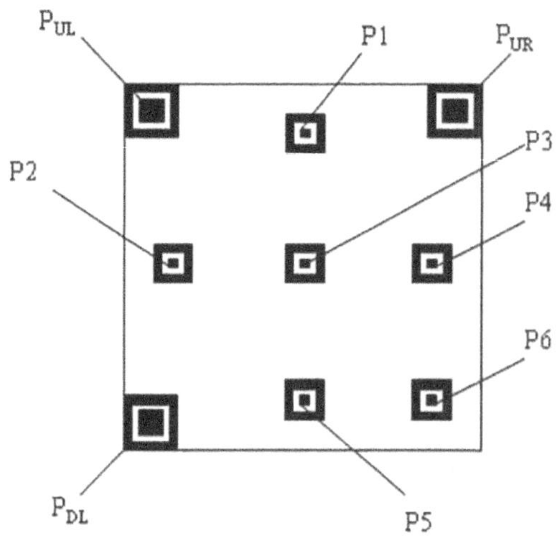

Figure 34 — Finder patterns and alignment patterns

1) Divide the pattern width W_{UL} of the upper left finder pattern P_{UL} by 7 to calculate the module size CP_{UL}.

$$CP_{UL} = W_{UL} / 7$$

2) Determine the provisional central coordinates of the alignment patterns P1 and P2 (see Figure 33), based on the coordinate of the center A of the upper left finder pattern P_{UL}, lines parallel to the guide lines AB and AC obtained in 7c), and the module size CP_{UL}.

3) Scan the outline of the white square in alignment pattern P1 and P2 starting from the pixel of the provisional central coordinate to find the actual central coordinates Xi and Yj (see Figure 35).

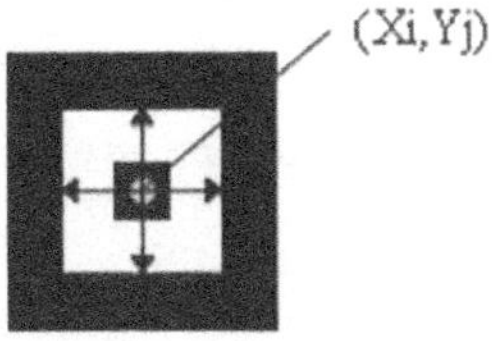

Figure 35 — Central coordinates of alignment pattern

4) Estimate the provisional central coordinate of the alignment pattern P3, based on the central coordinate of the upper left finder pattern P_{UL} and the actual central coordinates of the alignment patterns P1 and P2 obtained in step 3.

5) Find the actual central coordinate of the alignment pattern P3 by following the same procedure in step 3.

6) Find Lx, which is the center-to-center distance of the alignment patterns P2 and P3, and Ly, which is the center-to-center distance of the alignment patterns P1 and P3. Divide Lx and Ly by the defined spacing of the alignment patterns to obtain the module pitches CPx in the lower side and CPy in the right side in the upper left area of the symbol (see Figure 36).

$$CPx = Lx / AP$$

$$CPy = Ly / AP$$

where AP is the spacing in modules of the alignment pattern centers (see Table E.1).

In the same fashion, find Lx', which is the horizontal distance between the central coordinates of the upper left finder pattern P_{UL} and the central coordinates of the alignment pattern P1, and Ly', which is the vertical distance between the central coordinates of the upper left finder pattern P_{UL} and the central coordinates of the alignment pattern P2. Divide Lx' and Ly' by the formula below to obtain the module pitches CPx' in the upper side and CPy' in the left side in the upper left area of the symbol.

CPx' = Lx' / (Column coordinate of the central module of the alignment pattern P1

— Column coordinate of the central module of the upper left Finder Pattern P_{UL})

CPy' = Ly' / (Row coordinate of the central module of the alignment pattern P2

— Row coordinate of the central module of the upper left Finder Pattern P_{UL})

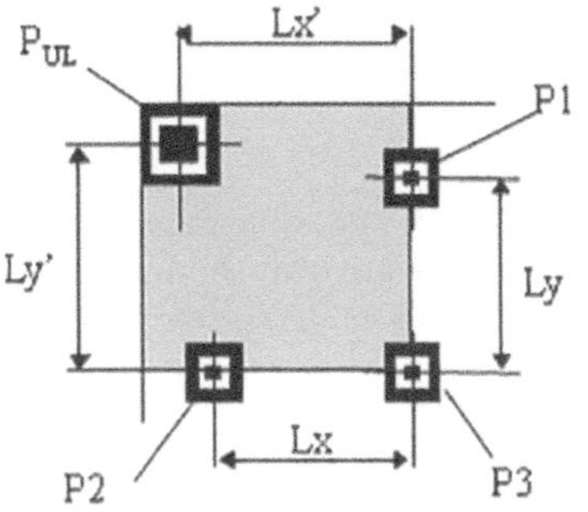

Figure 36 — Upper left area of symbol

7) Determine the sampling grid covering the upper left area of the symbol based on the module pitches CPx, CPx', CPy and CPy' representing each side in the upper left area of the symbol.

8) In the same fashion detemine the sampling grids for the upper right area (covered by the upper right finder pattern P_{UR}, alignment patterns P1, P3 and P4) and lower left area (covered by the lower left finder pattern P_{DL}, alignment patterns P2, P3 and P5) of the symbol.

9) For the alignment pattern P6 (see Figure 37), estimate its provisional central coordinate from the module pitches CPx' and CPy', the values of which are obtained from the spacings of alignment patterns P3, P4 and P5, guide lines passing through the centers of the alignment patterns P3 and P4, and P3 and P5 respectively, and the central coordinates of these Patterns.

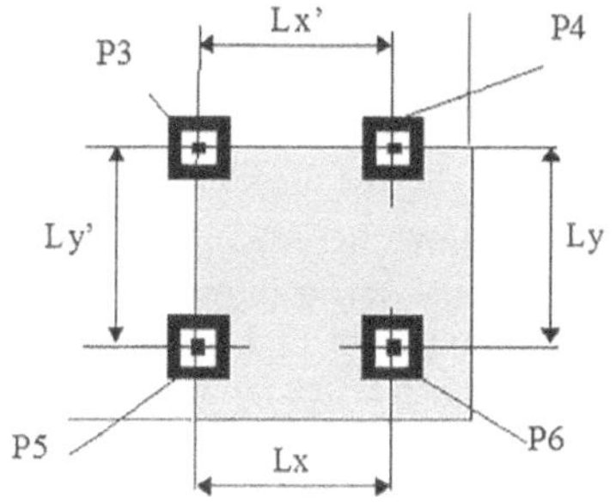

Figure 37 — Lower right area of symbol

10) Repeat steps 5) to 8) to determine the sampling grid for the lower right area of the symbol.

11) The same principles shall be applied to determine the sampling grids for any areas of the symbol not already covered.

i) Sample an area of 3 x 3 image pixels, centred on each intersection of the grid lines and determine whether it is dark or light based on the Global Threshold. Construct a bit matrix mapping the dark modules as binary 1 and light modules as binary 0.

j) Decode the format information adjacent to the upper left finder pattern as described in Annex C.3 to yield the Error Correction Level and the data mask pattern applied to the symbol. If errors exceeding the error correction capacity of the format information are detected, then follow the same procedure to decode the format information adjacent to the upper right and lower left finder patterns.

k) If a valid format information bit string cannot be derived, determine whether it is a valid sequence if read in the reverse direction and if so attempt to continue decoding as a mirror image symbol with the image row and column coordinates transposed.

l) Go to step y.

m) For Micro QR Code symbols, determine the possible angles of rotation of the symbol by analysing the angles of the lines from step b) 3) relative to the imaging sensor axes, as ϑ (see Figure 38), $\vartheta + 90°$, $\vartheta + 180°$ and $\vartheta + 270°$.

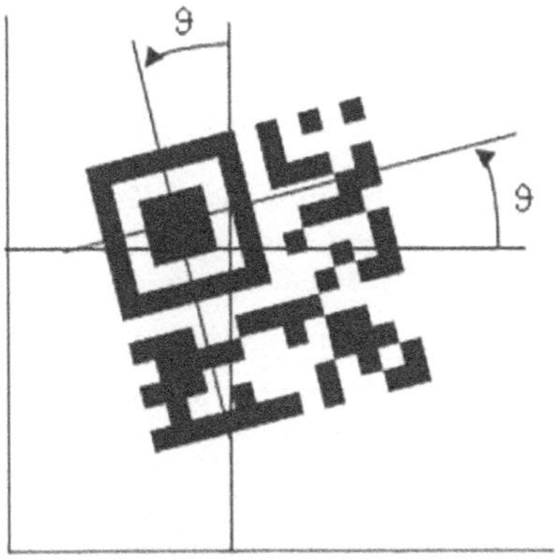

Figure 38 — angle ϑ relative to the imaging sensor axes

n) Plot three lines parallel to each axis of the finder pattern and equally spaced across the pattern and measure the distances from point A to point B on each line. The spacing is not limited but three lines shall be in the finder pattern.

o) Calculate the provisional module dimension X of the symbol in each axis as one seventh of the mean of the three distances A to B from step n.

p) Taking each side of the outer box of the finder pattern in turn, extend a line outward from the finder pattern in both directions, parallel to the edge and 0,5 X in from the edge.

q) Search for the timing patterns:

   1) Identify two edges of the finder pattern nominally perpendicular to each other, each of which has both

      i) a clear area of at least 1,5X in one direction;

      ii) alternating light and dark areas evenly spaced at 1X centres from the edge of the finder pattern in the opposite direction (a candidate timing pattern).

   2) Check that there is the same number of dark modules in each candidate timing pattern and that this number is between two and five.

r) Determine the provisional version of the symbol from the number of dark elements in the timing pattern:

   – If there are two dark elements, the symbol version is M1;

   – If there are three dark elements, the symbol version is M2;

   – If there are four dark elements, the symbol version is M3;

   – If there are five dark elements, the symbol version is M4.

s) From the centre of the first dark module in each side of the timing patterns extend a line parallel with the adjacent side of the finder pattern to intersect with the corresponding line from the other side and sample an area of 3 x 3 image pixels at 1X intervals along the line to determine the light or dark status of each module of the format information. Determine the format information bit string by taking the dark pixels as binary 1 and light pixels as binary 0.

t) Release masking of the format information by XORing the bit string with the pattern given in 7.9.2 and decode the format information (applying the error correction procedure given in Annex B if necessary) to yield the symbol number (and hence the version and error correction level of the symbol) and the data mask pattern applied to the symbol.

u) If the format information bit string is not a valid sequence, determine whether it is a valid sequence if read in the reverse direction and if so attempt to continue decoding as a mirror image symbol with the image row and column coordinates transposed. If no more than two bits differ from a valid sequence in Annex C substitute this sequence and decode the substituted format information to obtain the symbol number and the data mask pattern.

v) Confirm the module pitch X in each axis by dividing the overall width from the outer edge of the finder pattern adjacent to the quiet zone to the outer edge of the last dark module in the timing pattern by the number of modules corresponding to the symbol version.

w) Establish a sampling grid, corresponding to the version of the symbol, of lines spaced 1X apart in each axis, parallel to each other and to the side of the finder pattern, and running from the centres of the timing pattern modules and from similar positions in the finder pattern.

x) Sample an area of 3 x 3 image pixels, centred on each intersection of the grid lines, and determine whether it is dark or light based on the Global Threshold. Construct a bit matrix mapping the dark modules as binary 1 and light modules as binary 0.

y) XOR the data mask pattern with the encoding region of the symbol to release the data masking and restore the symbol characters representing data and error correction codewords. This reverses the effect of the data masking process applied during the encoding procedure.

z) Determine the symbol codewords in accordance with the placement rules in 7.7.3.

aa) Rearrange the codeword sequence into blocks as required for the symbol Version and Error Correction Level, by reversing the interleaving process defined in 7.6. step 3).

bb) Follow the error detection and correction decoding procedure in Annex B to correct errors and erasures up to the maximum correction capacity for the symbol version and Error Correction Level.

cc) Restore the original message bit stream by assembling the data blocks in sequence.

dd) Subdivide the data bit stream into segments each commencing with a mode indicator and the length of which is determined by the character count indicator following the mode indicator.

ee) Decode each segment according to the rules for the mode in force.

13 Autodiscrimination capability

QR Code can be used in an autodiscrimination environment with a number of other symbologies. (See Annex L). Although Model 1 and QR Code symbols can be autodiscriminated by analysis of the format information mask pattern, Model 1 symbols should not be used in the same environment as QR Code symbols.

14 Transmitted data

14.1 General principles

All encoded data characters shall be included in the data transmission. The function patterns, format and version information, error correction characters, Pad and Remainder characters shall not be transmitted. The default transmission mode for all data shall be as bytes.

The Structured Append header block shall not be transmitted by decoders operating in buffered mode which have reconstructed the complete message before transmission. If the decoder is operating in unbuffered mode the Structured Append header shall be transmitted as the first 2 bytes of every symbol.

More complex interpretations including the transmission of data in an Extended Channel Interpretation, are addressed below.

14.2 Symbology Identifier

ISO/IEC 15424 provides a standard procedure for reporting the symbology which has been read, together with options set in the decoder and any special features encountered in the symbol.

Once the structure of the data (including the use of any ECI) has been identified, the appropriate Symbology Identifier should be added by the decoder as a preamble to the transmitted data; if ECIs are used the Symbology Identifier is required. See Annex F for the Symbology Identifier and option values which apply to QR Code.

14.3 Extended Channel Interpretations

In systems where the ECI protocol is supported the transmission of the Symbology Identifier is required with every transmission. Whenever the ECI mode indicator is encountered, it shall be transmitted as the escape character $5C_{HEX}$, (which represents the backslash character "\" in ISO/IEC 8859-1 and in the AIM ECI specification and maps to the character "¥" in JIS X 0201). The codeword(s) representing the ECI Designator are converted into a 6 digit number by inverting the rules defined in Table 4. These 6 digits shall be transmitted as the corresponding 8-bit values in the range 30_{HEX} to 39_{HEX}, immediately following the escape character.

Application software recognizing **\nnnnnn** should interpret all subsequent characters as being from the ECI defined by the 6 digit designator. This interpretation remains in effect until:

- the end of the encoded data;
- a change to a new ECI signaled by mode indicator **0111**, subject to rules defined by the AIM ECI specification.

When reverting to the default interpretation the decoder shall output the appropriate escape sequence as prefix to the data.

If the character $5C_{HEX}$ needs to be used as encoded data, transmission shall be as follows: whenever character $5C_{HEX}$ occurs as data, two bytes of the value shall be transmitted, thus a single occurrence is always an escape character and a double occurrence indicates true data.

Example 1

a) Encoded data (hex):	41 42 43 5C 31 32 33 34
Transmitted data:	41 42 43 5C 5C 31 32 33 34
b) Encoded data:	ABC followed by <further data> encoded according to rules for ECI 123456.
Transmitted data:	41 42 43 5C 31 32 33 34 35 36 <further data>

Example 2 (using data in 7.4.2.2)

The message contains ECI mode indicator/ECI Designator/mode indicator/Character count indicator/Data in the form of

0111 00001001 0100 00000101 10100001 10100010 10100011 10100100 10100101

Symbology Identifier	**]Q2** (see Annex F) must be added to the data transmission.
Transmission (hex. values):	**5D 51 32 5C 30 30 30 30 30 39 A1 A2 A3 A4 A5**
Encoded data in ECI 000009:	**ΑΒΓΔΕ**

In Structured Append mode, when the ECI mode indicator is encountered at the beginning of the symbol, subsequent data characters shall be interpreted as being from the ECI(s) in force at the end of the preceding symbol.

NOTE 5C$_{HEX}$ is equivalent to the backslash character "\" in ISO/IEC 8859-1 and to "¥" in JIS X 0201.

14.4 FNC1

In the modes with implied FNC1 in either first or second position, this implied character cannot be transmitted directly as there is no byte value corresponding to it. It is therefore necessary to indicate its presence in the first or second position by the transmission of the relevant Symbology Identifier (**]Q3,]Q4,]Q5** or **]Q6** defined in Annex F shall be used). Elsewhere in these symbols it may occur in accordance with the relevant application specification as a data field separator, represented in Alphanumeric mode by the character **%** and in Byte mode by the character **GS** (ASCII/JIS8 value 1D$_{HEX}$). In both cases the decoder shall transmit ASCII/JIS8 value 1D$_{HEX}$.

If, in symbols in FNC1 mode, the character **%** needs to be encoded as data while in Alphanumeric mode, it shall be represented in the symbol by **%%**. If this is encountered the decoder shall transmit a single **%** character.

Annex A
(normative)

Error detection and correction generator polynomials

The check character generation polynomial is used to divide the data codeword polynomial, where each codeword is the coefficient of the dividend polynomial in descending power order. The coefficients of the remainder of this division are the error correction codeword values.

Table A.1 shows the generator polynomials for the error correction codes which are used for each Version and Level, for all QR Code symbols. The number of error correction codewords required for a particular version and error correction level can be obtained from Table 9. In the table, α is the primitive element 2 under GF(2^8). Each generator polynomial is the product of the first degree polynomials: $x-2^0$, $x-2^1$, ..., $x-2^{n-1}$; where n is the degree of the generator polynomial.

Table A.1 — Generator polynomials for Reed-Solomon error correction codewords

Number of error correction code-words	Generator polynomials
2	$x^2 + \alpha^{25}x + \alpha$
5	$x^5 + \alpha^{113}x^4 + \alpha^{164}x^3 + \alpha^{166}x^2 + \alpha^{119}x + \alpha^{10}$
6	$x^6 + \alpha^{166}x^5 + x^4 + \alpha^{134}x^3 + \alpha^5x^2 + \alpha^{176}x + \alpha^{15}$
7	$x^7 + \alpha^{87}x^6 + \alpha^{229}x^5 + \alpha^{146}x^4 + \alpha^{149}x^3 + \alpha^{238}x^2 + \alpha^{102}x + \alpha^{21}$
8	$x^8 + \alpha^{175}x^7 + \alpha^{238}x^6 + \alpha^{208}x^5 + \alpha^{249}x^4 + \alpha^{215}x^3 + \alpha^{252}x^2 + \alpha^{196}x + \alpha^{28}$
10	$x^{10} + \alpha^{251}x^9 + \alpha^{67}x^8 + \alpha^{46}x^7 + \alpha^{61}x^6 + \alpha^{118}x^5 + \alpha^{70}x^4 + \alpha^{64}x^3 + \alpha^{94}x^2 + \alpha^{32}x + \alpha^{45}$
13	$x^{13} + \alpha^{74}x^{12} + \alpha^{152}x^{11} + \alpha^{176}x^{10} + \alpha^{100}x^9 + \alpha^{86}x^8 + \alpha^{100}x^7 + \alpha^{106}x^6 + \alpha^{104}x^5 + \alpha^{130}x^4 + \alpha^{218}x^3 + \alpha^{206}x^2 + \alpha^{140}x + \alpha^{78}$
14	$x^{14} + \alpha^{199}x^{13} + \alpha^{249}x^{12} + \alpha^{155}x^{11} + \alpha^{48}x^{10} + \alpha^{190}x^9 + \alpha^{124}x^8 + \alpha^{218}x^7 + \alpha^{137}x^6 + \alpha^{216}x^5 + \alpha^{87}x^4 + \alpha^{207}x^3 + \alpha^{59}x^2 + \alpha^{22}x + \alpha^{91}$
15	$x^{15} + \alpha^8x^{14} + \alpha^{183}x^{13} + \alpha^{61}x^{12} + \alpha^{91}x^{11} + \alpha^{202}x^{10} + \alpha^{37}x^9 + \alpha^{51}x^8 + \alpha^{58}x^7 + \alpha^{58}x^6 + \alpha^{237}x^5 + \alpha^{140}x^4 + \alpha^{124}x^3 + \alpha^5x^2 + \alpha^{99}x + \alpha^{105}$
16	$x^{16} + \alpha^{120}x^{15} + \alpha^{104}x^{14} + \alpha^{107}x^{13} + \alpha^{109}x^{12} + \alpha^{102}x^{11} + \alpha^{161}x^{10} + \alpha^{76}x^9 + \alpha^3x^8 + \alpha^{91}x^7 + \alpha^{191}x^6 + \alpha^{147}x^5 + \alpha^{169}x^4\alpha^{182}x^3 + \alpha^{194}x^2 + \alpha^{225}x + \alpha^{120}$
17	$x^{17} + \alpha^{43}x^{16} + \alpha^{139}x^{15} + \alpha^{206}x^{14} + \alpha^{78}x^{13} + \alpha^{43}x^{12} + \alpha^{239}x^{11} + \alpha^{123}x^{10} + \alpha^{206}x^9 + \alpha^{214}x^8 + \alpha^{147}x^7 + \alpha^{24}x^6 + \alpha^{99}x^5 + \alpha^{150}x^4 + \alpha^{39}x^3 + \alpha^{243}x^2 + \alpha^{163}x + \alpha^{136}$
18	$x^{18} + \alpha^{215}x^{17} + \alpha^{234}x^{16} + \alpha^{158}x^{15} + \alpha^{94}x^{14} + \alpha^{184}x^{13} + \alpha^{97}x^{12} + \alpha^{118}x^{11} + \alpha^{170}x^{10} + \alpha^{79}x^9 + \alpha^{187}x^8 + \alpha^{152}x^7 + \alpha^{148}x^6 + \alpha^{252}x^5 + \alpha^{179}x^4 + \alpha^5x^3 + \alpha^{98}x^2 + \alpha^{96}x + \alpha^{153}$
20	$x^{20} + \alpha^{17}x^{19} + \alpha^{60}x^{18} + \alpha^{79}x^{17} + \alpha^{50}x^{16} + \alpha^{61}x^{15} + \alpha^{163}x^{14} + \alpha^{26}x^{13} + \alpha^{187}x^{12} + \alpha^{202}x^{11} + \alpha^{180}x^{10} + \alpha^{221}x^9 + \alpha^{225}x^8 + \alpha^{83}x^7 + \alpha^{239}x^6 + \alpha^{156}x^5 + \alpha^{164}x^4 + \alpha^{212}x^3 + \alpha^{212}x^2 + \alpha^{188}x + \alpha^{190}$
22	$x^{22} + \alpha^{210}x^{21} + \alpha^{171}x^{20} + \alpha^{247}x^{19} + \alpha^{242}x^{18} + \alpha^{93}x^{17} + \alpha^{230}x^{16} + \alpha^{14}x^{15} + \alpha^{109}x^{14} + \alpha^{221}x^{13} + \alpha^{53}x^{12} + \alpha^{200}x^{11} + \alpha^{74}x^{10} + \alpha^8x^9 + \alpha^{172}x^8 + \alpha^{98}x^7 + \alpha^{80}x^6 + \alpha^{219}x^5 + \alpha^{134}x^4 + \alpha^{160}x^3 + \alpha^{105}x^2 + \alpha^{165}x + \alpha^{231}$

Table A.1 *(continued)*

Number of error correction code-words	Generator polynomials
24	$x^{24} + \alpha^{229}x^{23} + \alpha^{121}x^{22} + \alpha^{135}x^{21} + \alpha^{48}x^{20} + \alpha^{211}x^{19} + \alpha^{117}x^{18} + \alpha^{251}x^{17} + \alpha^{126}x^{16} + \alpha^{159}x^{15} + \alpha^{180}x^{14} + \alpha^{169}x^{13} + \alpha^{152}x^{12} + \alpha^{192}x^{11} + \alpha^{226}x^{10} + \alpha^{228}x^{9} + \alpha^{218}x^{8} + \alpha^{111}x^{7} + x^{6} + \alpha^{117}x^{5} + \alpha^{232}x^{4} + \alpha^{87}x^{3} + \alpha^{96}x^{2} + \alpha^{227}x + \alpha^{21}$
26	$x^{26} + \alpha^{173}x^{25} + \alpha^{125}x^{24} + \alpha^{158}x^{23} + \alpha^{2}x^{22} + \alpha^{103}x^{21} + \alpha^{182}x^{20} + \alpha^{118}x^{19} + \alpha^{17}x^{18} + \alpha^{145}x^{17} + \alpha^{201}x^{16} + \alpha^{111}x^{15} + \alpha^{28}x^{14} + \alpha^{165}x^{13} + \alpha^{53}x^{12} + \alpha^{161}x^{11} + \alpha^{21}x^{10} + \alpha^{245}x^{9} + \alpha^{142}x^{8} + \alpha^{13}x^{7} + \alpha^{102}x^{6} + \alpha^{48}x^{5} + \alpha^{227}x^{4} + \alpha^{153}x^{3} + \alpha^{145}x^{2} + \alpha^{218}x + \alpha^{70}$
28	$x^{28} + \alpha^{168}x^{27} + \alpha^{223}x^{26} + \alpha^{200}x^{25} + \alpha^{104}x^{24} + \alpha^{224}x^{23} + \alpha^{234}x^{22} + \alpha^{108}x^{21} + \alpha^{180}x^{20} + \alpha^{110}x^{19} + \alpha^{190}x^{18} + \alpha^{195}x^{17} + \alpha^{147}x^{16} + \alpha^{205}x^{15} + \alpha^{27}x^{14} + \alpha^{232}x^{13} + \alpha^{201}x^{12} + \alpha^{21}x^{11} + \alpha^{43}x^{10} + \alpha^{245}x^{9} + \alpha^{87}x^{8} + \alpha^{42}x^{7} + \alpha^{195}x^{6} + \alpha^{212}x^{5} + \alpha^{119}x^{4} + \alpha^{242}x^{3} + \alpha^{37}x^{2} + \alpha^{9}x + \alpha^{123}$
30	$x^{30} + \alpha^{41}x^{29} + \alpha^{173}x^{28} + \alpha^{145}x^{27} + \alpha^{152}x^{26} + \alpha^{216}x^{25} + \alpha^{31}x^{24} + \alpha^{179}x^{23} + \alpha^{182}x^{22} + \alpha^{50}x^{21} + \alpha^{48}x^{20} + \alpha^{110}x^{19} + \alpha^{86}x^{18} + \alpha^{239}x^{17} + \alpha^{96}x^{16} + \alpha^{222}x^{15} + \alpha^{125}x^{14} + \alpha^{42}x^{13} + \alpha^{173}x^{12} + \alpha^{226}x^{11} + \alpha^{193}x^{10} + \alpha^{224}x^{9} + \alpha^{130}x^{8} + \alpha^{156}x^{7} + \alpha^{37}x^{6} + \alpha^{251}x^{5} + \alpha^{216}x^{4} + \alpha^{238}x^{3} + \alpha^{40}x^{2} + \alpha^{192}x + \alpha^{180}$
32	$x^{32}\alpha^{10}x^{31} + \alpha^{6}x^{30} + \alpha^{106}x^{29} + \alpha^{190}x^{28} + \alpha^{249}x^{27} + \alpha^{167}x^{26} + \alpha^{4}x^{25} + \alpha^{67}x^{24} + \alpha^{209}x^{23} + \alpha^{138}x^{22} + \alpha^{138}x^{21} + \alpha^{32}x^{20} + \alpha^{242}x^{19} + \alpha^{123}x^{18} + \alpha^{89}x^{17} + \alpha^{27}x^{16} + \alpha^{120}x^{15} + \alpha^{185}x^{14} + \alpha^{80}x^{13} + \alpha^{156}x^{12} + \alpha^{38}x^{11} + \alpha^{69}x^{10} + \alpha^{171}x^{9} + \alpha^{60}x^{8} + \alpha^{28}x^{7} + \alpha^{222}x^{6} + \alpha^{80}x^{5} + \alpha^{52}x^{4} + \alpha^{254}x^{3} + \alpha^{185}x^{2} + \alpha^{220}x + \alpha^{241}$
34	$x^{34} + \alpha^{111}x^{33} + \alpha^{77}x^{32} + \alpha^{146}x^{31} + \alpha^{94}x^{30} + \alpha^{26}x^{29} + \alpha^{21}x^{28} + \alpha^{108}x^{27} + \alpha^{19}x^{26} + \alpha^{105}x^{25} + \alpha^{94}x^{24} + \alpha^{113}x^{23} + \alpha^{193}x^{22} + \alpha^{86}x^{21} + \alpha^{140}x^{20} + \alpha^{163}x^{19} + \alpha^{125}x^{18} + \alpha^{58}x^{17} + \alpha^{158}x^{16} + \alpha^{229}x^{15} + \alpha^{239}x^{14} + \alpha^{218}x^{13} + \alpha^{103}x^{12} + \alpha^{56}x^{11} + \alpha^{70}x^{10} + \alpha^{114}x^{9} + \alpha^{61}x^{8} + \alpha^{183}x^{7} + \alpha^{129}x^{6} + \alpha^{167}x^{5} + \alpha^{13}x^{4} + \alpha^{98}x^{3} + \alpha^{62}x^{2} + \alpha^{129}x + \alpha^{51}$
36	$x^{36} + \alpha^{200}x^{35} + \alpha^{183}x^{34} + \alpha^{98}x^{33} + \alpha^{16}x^{32} + \alpha^{172}x^{31} + \alpha^{31}x^{30} + \alpha^{246}x^{29} + \alpha^{234}x^{28} + \alpha^{60}x^{27} + \alpha^{152}x^{26} + \alpha^{115}x^{25} + x^{24} + \alpha^{167}x^{23} + \alpha^{152}x^{22} + \alpha^{113}x^{21} + \alpha^{248}x^{20} + \alpha^{238}x^{19} + \alpha^{107}x^{18} + \alpha^{18}x^{17} + \alpha^{63}x^{16} + \alpha^{218}x^{15} + \alpha^{37}x^{14} + \alpha^{87}x^{13} + \alpha^{210}x^{12} + \alpha^{105}x^{11} + \alpha^{177}x^{10} + \alpha^{120}x^{9} + \alpha^{74}x^{8} + \alpha^{121}x^{7} + \alpha^{196}x^{6} + \alpha^{117}x^{5} + \alpha^{251}x^{4} + \alpha^{113}x^{3} + \alpha^{233}x^{2} + \alpha^{30}x + \alpha^{120}$
40	$x^{40} + \alpha^{59}x^{39} + \alpha^{116}x^{38} + \alpha^{79}x^{37} + \alpha^{161}x^{36} + \alpha^{252}x^{35} + \alpha^{98}x^{34} + \alpha^{128}x^{33} + \alpha^{205}x^{32} + \alpha^{128}x^{31} + \alpha^{161}x^{30} + \alpha^{247}x^{29} + \alpha^{57}x^{28} + \alpha^{163}x^{27} + \alpha^{56}x^{26} + \alpha^{235}x^{25} + \alpha^{106}x^{24} + \alpha^{53}x^{23} + \alpha^{26}x^{22} + \alpha^{187}x^{21} + \alpha^{174}x^{20} + \alpha^{226}x^{19} + \alpha^{104}x^{18} + \alpha^{170}x^{17} + \alpha^{7}x^{16} + \alpha^{175}x^{15} + \alpha^{35}x^{14} + \alpha^{181}x^{13} + \alpha^{114}x^{12} + \alpha^{88}x^{11} + \alpha^{41}x^{10} + \alpha^{47}x^{9} + \alpha^{163}x^{8} + \alpha^{125}x^{7} + \alpha^{134}x^{6} + \alpha^{72}x^{5} + \alpha^{20}x^{4} + \alpha^{232}x^{3} + \alpha^{53}x^{2} + \alpha^{35}x + \alpha^{15}$
42	$x^{42} + \alpha^{250}x^{41} + \alpha^{103}x^{40} + \alpha^{221}x^{39} + \alpha^{230}x^{38} + \alpha^{25}x^{37} + \alpha^{18}x^{36} + \alpha^{137}x^{35} + \alpha^{231}x^{34} + x^{33} + \alpha^{3}x^{32} + \alpha^{58}x^{31} + \alpha^{242}x^{30} + \alpha^{221}x^{29} + \alpha^{191}x^{28} + \alpha^{110}x^{27} + \alpha^{84}x^{26} + \alpha^{230}x^{25} + \alpha^{8}x^{24} + \alpha^{188}x^{23} + \alpha^{106}x^{22} + \alpha^{96}x^{21} + \alpha^{147}x^{20} + \alpha^{15}x^{19} + \alpha^{131}x^{18} + \alpha^{139}x^{17} + \alpha^{34}x^{16} + \alpha^{101}x^{15} + \alpha^{223}x^{14} + \alpha^{39}x^{13} + \alpha^{101}x^{12} + \alpha^{213}x^{11} + \alpha^{199}x^{10} + \alpha^{237}x^{9} + \alpha^{254}x^{8} + \alpha^{201}x^{7} + \alpha^{123}x^{6} + \alpha^{171}x^{5} + \alpha^{162}x^{4} + \alpha^{194}x^{3} + \alpha^{117}x^{2} + \alpha^{50}x + \alpha^{96}$
44	$x^{44} + \alpha^{190}x^{43} + \alpha^{7}x^{42} + \alpha^{61}x^{41} + \alpha^{121}x^{40} + \alpha^{71}x^{39} + \alpha^{246}x^{38} + \alpha^{69}x^{37} + \alpha^{55}x^{36} + \alpha^{168}x^{35} + \alpha^{188}x^{34} + \alpha^{89}x^{33} + \alpha^{243}x^{32} + \alpha^{191}x^{31} + \alpha^{25}x^{30} + \alpha^{72}x^{29} + \alpha^{123}x^{28} + \alpha^{9}x^{27} + \alpha^{145}x^{26} + \alpha^{14}x^{25} + \alpha^{247}x^{24} + \alpha x^{23} + \alpha^{238}x^{22} + \alpha^{44}x^{21} + \alpha^{78}x^{20} + \alpha^{143}x^{19} + \alpha^{62}x^{18} + \alpha^{224}x^{17} + \alpha^{126}x^{16} + \alpha^{118}x^{15} + \alpha^{114}x^{14} + \alpha^{68}x^{13} + \alpha^{163}x^{12} + \alpha^{52}x^{11} + \alpha^{194}x^{10} + \alpha^{217}x^{9} + \alpha^{147}x^{8} + \alpha^{204}x^{7} + \alpha^{169}x^{6} + \alpha^{37}x^{5} + \alpha^{130}x^{4} + \alpha^{113}x^{3} + \alpha^{102}x^{2} + \alpha^{73}x + \alpha^{181}$

Table A.1 *(continued)*

Number of error correction code-words	Generator polynomials
46	$x^{46} + \alpha^{112}x^{45} + \alpha^{94}x^{44} + \alpha^{88}x^{43} + \alpha^{112}x^{42} + \alpha^{253}x^{41} + \alpha^{224}x^{40} + \alpha^{202}x^{39} + \alpha^{115}x^{38} + \alpha^{187}x^{37} + \alpha^{99}x^{36} + \alpha^{89}x^{35} + \alpha^{5}x^{34} + \alpha^{54}x^{33} + \alpha^{113}x^{32} + \alpha^{129}x^{31} + \alpha^{44}x^{30} + \alpha^{58}x^{29} + \alpha^{16}x^{28} + \alpha^{135}x^{27} + \alpha^{216}x^{26} + \alpha^{169}x^{25} + \alpha^{211}x^{24} + \alpha^{36}x^{23} + \alpha x^{22} + \alpha^{4}x^{21} + \alpha^{96}x^{20} + \alpha^{60}x^{19} + \alpha^{241}x^{18} + \alpha^{73}x^{17} + \alpha^{104}x^{16} + \alpha^{234}x^{15} + \alpha^{8}x^{14} + \alpha^{249}x^{13} + \alpha^{245}x^{12} + \alpha^{119}x^{11} + \alpha^{174}x^{10} + \alpha^{52}x^{9} + \alpha^{25}x^{8} + \alpha^{157}x^{7} + \alpha^{224}x^{6} + \alpha^{43}x^{5} + \alpha^{202}x^{4} + \alpha^{223}x^{3} + \alpha^{19}x^{2} + \alpha^{82}x + \alpha^{15}$
48	$x^{48} + \alpha^{228}x^{47} + \alpha^{25}x^{46} + \alpha^{196}x^{45} + \alpha^{130}x^{44} + \alpha^{211}x^{43} + \alpha^{146}x^{42} + \alpha^{60}x^{41} + \alpha^{24}x^{40} + \alpha^{251}x^{39} + \alpha^{90}x^{38} + \alpha^{39}x^{37} + \alpha^{102}x^{36} + \alpha^{240}x^{35} + \alpha^{61}x^{34} + \alpha^{178}x^{33} + \alpha^{63}x^{32} + \alpha^{46}x^{31} + \alpha^{123}x^{30} + \alpha^{115}x^{29} + \alpha^{18}x^{28} + \alpha^{221}x^{27} + \alpha^{111}x^{26} + \alpha^{135}x^{25} + \alpha^{160}x^{24} + \alpha^{182}x^{23} + \alpha^{205}x^{22} + \alpha^{107}x^{21} + \alpha^{206}x^{20} + \alpha^{95}x^{19} + \alpha^{150}x^{18} + \alpha^{120}x^{17} + \alpha^{184}x^{16} + \alpha^{91}x^{15} + \alpha^{21}x^{14} + \alpha^{247}x^{13} + \alpha^{156}x^{12} + \alpha^{140}x^{11} + \alpha^{238}x^{10} + \alpha^{191}x^{9} + \alpha^{11}x^{8} + \alpha^{94}x^{7} + \alpha^{227}x^{6} + \alpha^{84}x^{5} + \alpha^{50}x^{4} + \alpha^{163}x^{3} + \alpha^{39}x^{2} + \alpha^{34}x + \alpha^{108}$
50	$x^{50} + \alpha^{232}x^{49} + \alpha^{125}x^{48} + \alpha^{157}x^{47} + \alpha^{161}x^{46} + \alpha^{164}x^{45} + \alpha^{9}x^{44} + \alpha^{118}x^{43} + \alpha^{46}x^{42} + \alpha^{209}x^{41} + \alpha^{99}x^{40} + \alpha^{203}x^{39} + \alpha^{193}x^{38} + \alpha^{35}x^{37} + \alpha^{3}x^{36} + \alpha^{209}x^{35} + \alpha^{111}x^{34} + \alpha^{195}x^{33} + \alpha^{242}x^{32} + \alpha^{203}x^{31} + \alpha^{225}x^{30} + \alpha^{46}x^{29} + \alpha^{13}x^{28} + \alpha^{32}x^{27} + \alpha^{160}x^{26} + \alpha^{126}x^{25} + \alpha^{209}x^{24} + \alpha^{130}x^{23} + \alpha^{160}x^{22} + \alpha^{242}x^{21} + \alpha^{215}x^{20} + \alpha^{242}x^{19} + \alpha^{75}x^{18} + \alpha^{77}x^{17} + \alpha^{42}x^{16} + \alpha^{189}x^{15} + \alpha^{32}x^{14} + \alpha^{113}x^{13} + \alpha^{65}x^{12} + \alpha^{124}x^{11} + \alpha^{69}x^{10} + \alpha^{228}x^{9} + \alpha^{114}x^{8} + \alpha^{235}x^{7} + \alpha^{175}x^{6} + \alpha^{124}x^{5} + \alpha^{170}x^{4} + \alpha^{215}x^{3} + \alpha^{232}x^{2} + \alpha^{133}x + \alpha^{205}$
52	$x^{52} + \alpha^{116}x^{51} + \alpha^{50}x^{50} + \alpha^{86}x^{49} + \alpha^{186}x^{48} + \alpha^{50}x^{47} + \alpha^{220}x^{46} + \alpha^{251}x^{45} + \alpha^{89}x^{44} + \alpha^{192}x^{43} + \alpha^{46}x^{42} + \alpha^{86}x^{41} + \alpha^{127}x^{40} + \alpha^{124}x^{39} + \alpha^{19}x^{38} + \alpha^{184}x^{37} + \alpha^{233}x^{36} + \alpha^{151}x^{35} + \alpha^{215}x^{34} + \alpha^{22}x^{33} + \alpha^{14}x^{32} + \alpha^{59}x^{31} + \alpha^{145}x^{30} + \alpha^{37}x^{29} + \alpha^{242}x^{28} + \alpha^{203}x^{27} + \alpha^{134}x^{26} + \alpha^{254}x^{25} + \alpha^{89}x^{24} + \alpha^{190}x^{23} + \alpha^{94}x^{22} + \alpha^{59}x^{21} + \alpha^{65}x^{20} + \alpha^{124}x^{19} + \alpha^{113}x^{18} + \alpha^{100}x^{17} + \alpha^{233}x^{16} + \alpha^{235}x^{15} + \alpha^{121}x^{14} + \alpha^{22}x^{13} + \alpha^{76}x^{12} + \alpha^{86}x^{11} + \alpha^{97}x^{10} + \alpha^{39}x^{9} + \alpha^{242}x^{8} + \alpha^{200}x^{7} + \alpha^{220}x^{6} + \alpha^{101}x^{5} + \alpha^{33}x^{4} + \alpha^{239}x^{3} + \alpha^{254}x^{2} + \alpha^{116}x + \alpha^{51}$
54	$x^{54} + \alpha^{183}x^{53} + \alpha^{26}x^{52} + \alpha^{201}x^{51} + \alpha^{87}x^{50} + \alpha^{210}x^{49} + \alpha^{221}x^{48} + \alpha^{113}x^{47} + \alpha^{21}x^{46} + \alpha^{46}x^{45} + \alpha^{65}x^{44} + \alpha^{45}x^{43} + \alpha^{50}x^{42} + \alpha^{238}x^{41} + \alpha^{184}x^{40} + \alpha^{249}x^{39} + \alpha^{225}x^{38} + \alpha^{102}x^{37} + \alpha^{58}x^{36} + \alpha^{209}x^{35} + \alpha^{218}x^{34} + \alpha^{109}x^{33} + \alpha^{165}x^{32} + \alpha^{26}x^{31} + \alpha^{95}x^{30} + \alpha^{184}x^{29} + \alpha^{192}x^{28} + \alpha^{52}x^{27} + \alpha^{245}x^{26} + \alpha^{35}x^{25} + \alpha^{254}x^{24} + \alpha^{238}x^{23} + \alpha^{175}x^{22} + \alpha^{172}x^{21} + \alpha^{79}x^{20} + \alpha^{123}x^{19} + \alpha^{25}x^{18} + \alpha^{122}x^{17} + \alpha^{43}x^{16} + \alpha^{120}x^{15} + \alpha^{108}x^{14} + \alpha^{215}x^{13} + \alpha^{80}x^{12} + \alpha^{128}x^{11} + \alpha^{201}x^{10} + \alpha^{235}x^{9} + \alpha^{8}x^{8} + \alpha^{153}x^{7} + \alpha^{59}x^{6} + \alpha^{101}x^{5} + \alpha^{31}x^{4} + \alpha^{198}x^{3} + \alpha^{76}x^{2} + \alpha^{31}x + \alpha^{156}$
56	$x^{56} + \alpha^{106}x^{55} + \alpha^{120}x^{54} + \alpha^{107}x^{53} + \alpha^{157}x^{52} + \alpha^{164}x^{51} + \alpha^{216}x^{50} + \alpha^{112}x^{49} + \alpha^{116}x^{48} + \alpha^{2}x^{47} + \alpha^{91}x^{46} + \alpha^{248}x^{45} + \alpha^{163}x^{44} + \alpha^{36}x^{43} + \alpha^{201}x^{42} + \alpha^{202}x^{41} + \alpha^{229}x^{40} + \alpha^{6}x^{39} + \alpha^{144}x^{38} + \alpha^{254}x^{37} + \alpha^{155}x^{36} + \alpha^{135}x^{35} + \alpha^{208}x^{34} + \alpha^{170}x^{33} + \alpha^{209}x^{32} + \alpha^{12}x^{31} + \alpha^{139}x^{30} + \alpha^{127}x^{29} + \alpha^{142}x^{28} + \alpha^{182}x^{27} + \alpha^{249}x^{26} + \alpha^{177}x^{25} + \alpha^{174}x^{24} + \alpha^{190}x^{23} + \alpha^{28}x^{22} + \alpha^{10}x^{21} + \alpha^{85}x^{20} + \alpha^{239}x^{19} + \alpha^{184}x^{18} + \alpha^{101}x^{17} + \alpha^{124}x^{16} + \alpha^{152}x^{15} + \alpha^{206}x^{14} + \alpha^{96}x^{13} + \alpha^{23}x^{12} + \alpha^{163}x^{11} + \alpha^{61}x^{10} + \alpha^{27}x^{9} + \alpha^{196}x^{8} + \alpha^{247}x^{7} + \alpha^{151}x^{6} + \alpha^{154}x^{5} + \alpha^{202}x^{4} + \alpha^{207}x^{3} + \alpha^{20}x^{2} + \alpha^{61}x + \alpha^{10}$
58	$x^{58} + \alpha^{82}x^{57} + \alpha^{116}x^{56} + \alpha^{26}x^{55} + \alpha^{247}x^{54} + \alpha^{66}x^{53} + \alpha^{27}x^{52} + \alpha^{62}x^{51} + \alpha^{107}x^{50} + \alpha^{252}x^{49} + \alpha^{182}x^{48} + \alpha^{200}x^{47} + \alpha^{185}x^{46} + \alpha^{235}x^{45} + \alpha^{55}x^{44} + \alpha^{251}x^{43} + \alpha^{242}x^{42} + \alpha^{210}x^{41} + \alpha^{144}x^{40} + \alpha^{154}x^{39} + \alpha^{237}x^{38} + \alpha^{176}x^{37} + \alpha^{141}x^{36} + \alpha^{192}x^{35} + \alpha^{248}x^{34} + \alpha^{152}x^{33} + \alpha^{249}x^{32} + \alpha^{206}x^{31} + \alpha^{85}x^{30} + \alpha^{253}x^{29} + \alpha^{142}x^{28} + \alpha^{65}x^{27} + \alpha^{165}x^{26} + \alpha^{125}x^{25} + \alpha^{23}x^{24} + \alpha^{24}x^{23} + \alpha^{30}x^{22} + \alpha^{122}x^{21} + \alpha^{240}x^{20} + \alpha^{214}x^{19} + \alpha^{6}x^{18} + \alpha^{129}x^{17} + \alpha^{218}x^{16} + \alpha^{29}x^{15} + \alpha^{145}x^{14} + \alpha^{127}x^{13} + \alpha^{134}x^{12} + \alpha^{206}x^{11} + \alpha^{245}x^{10} + \alpha^{117}x^{9} + \alpha^{29}x^{8} + \alpha^{41}x^{7} + \alpha^{63}x^{6} + \alpha^{159}x^{5} + \alpha^{142}x^{4} + \alpha^{233}x^{3} + \alpha^{125}x^{2} + \alpha^{148}x + \alpha^{123}$

Table A.1 *(continued)*

Number of error correction code-words	Generator polynomials
60	$x^{60} + \alpha^{107}x^{59} + \alpha^{140}x^{58} + \alpha^{26}x^{57} + \alpha^{12}x^{56} + \alpha^{9}x^{55} + \alpha^{141}x^{54} + \alpha^{243}x^{53} + \alpha^{197}x^{52} + \alpha^{226}x^{51} + \alpha^{197}x^{50} + \alpha^{219}x^{49} + \alpha^{45}x^{48} + \alpha^{211}x^{47} + \alpha^{101}x^{46} + \alpha^{219}x^{45} + \alpha^{120}x^{44} + \alpha^{28}x^{43} + \alpha^{181}x^{42} + \alpha^{127}x^{41} + \alpha^{6}x^{40} + \alpha^{100}x^{39} + \alpha^{247}x^{38} + \alpha^{2}x^{37} + \alpha^{205}x^{36} + \alpha^{198}x^{35} + \alpha^{57}x^{34} + \alpha^{115}x^{33} + \alpha^{219}x^{32} + \alpha^{101}x^{31} + \alpha^{109}x^{30} + \alpha^{160}x^{29} + \alpha^{82}x^{28} + \alpha^{37}x^{27} + \alpha^{38}x^{26} + \alpha^{238}x^{25} + \alpha^{49}x^{24} + \alpha^{160}x^{23} + \alpha^{209}x^{22} + \alpha^{121}x^{21} + \alpha^{86}x^{20} + \alpha^{11}x^{19} + \alpha^{124}x^{18} + \alpha^{30}x^{17} + \alpha^{181}x^{16} + \alpha^{84}x^{15} + \alpha^{25}x^{14} + \alpha^{194}x^{13} + \alpha^{87}x^{12} + \alpha^{65}x^{11} + \alpha^{102}x^{10} + \alpha^{190}x^{9} + \alpha^{220}x^{8} + \alpha^{70}x^{7} + \alpha^{27}x^{6} + \alpha^{209}x^{5} + \alpha^{16}x^{4} + \alpha^{89}x^{3} + \alpha^{7}x^{2} + \alpha^{33}x + \alpha^{240}$
62	$x^{62} + \alpha^{65}x^{61} + \alpha^{202}x^{60} + \alpha^{113}x^{59} + \alpha^{98}x^{58} + \alpha^{71}x^{57} + \alpha^{223}x^{56} + \alpha^{248}x^{55} + \alpha^{118}x^{54} + \alpha^{214}x^{53} + \alpha^{94}x^{52} + x^{51} + \alpha^{122}x^{50} + \alpha^{37}x^{49} + \alpha^{23}x^{48} + \alpha^{2}x^{47} + \alpha^{228}x^{46} + \alpha^{58}x^{45} + \alpha^{121}x^{44} + \alpha^{7}x^{43} + \alpha^{105}x^{42} + \alpha^{135}x^{41} + \alpha^{78}x^{40} + \alpha^{243}x^{39} + \alpha^{118}x^{38} + \alpha^{70}x^{37} + \alpha^{76}x^{36} + \alpha^{223}x^{35} + \alpha^{89}x^{34} + \alpha^{72}x^{33} + \alpha^{50}x^{32} + \alpha^{70}x^{31} + \alpha^{111}x^{30} + \alpha^{194}x^{29} + \alpha^{17}x^{28} + \alpha^{212}x^{27} + \alpha^{126}x^{26} + \alpha^{181}x^{25} + \alpha^{35}x^{24} + \alpha^{221}x^{23} + \alpha^{117}x^{22} + \alpha^{235}x^{21} + \alpha^{11}x^{20} + \alpha^{229}x^{19} + \alpha^{149}x^{18} + \alpha^{147}x^{17} + \alpha^{123}x^{16} + \alpha^{213}x^{15} + \alpha^{40}x^{14} + \alpha^{115}x^{13} + \alpha^{6}x^{12} + \alpha^{200}x^{11} + \alpha^{100}x^{10} + \alpha^{26}x^{9} + \alpha^{246}x^{8} + \alpha^{182}x^{7} + \alpha^{218}x^{6} + \alpha^{127}x^{5} + \alpha^{215}x^{4} + \alpha^{36}x^{3} + \alpha^{186}x^{2} + \alpha^{110}x + \alpha^{106}$
64	$x^{64} + \alpha^{45}x^{63} + \alpha^{51}x^{62} + \alpha^{175}x^{61} + \alpha^{9}x^{60} + \alpha^{7}x^{59} + \alpha^{158}x^{58} + \alpha^{159}x^{57} + \alpha^{49}x^{56} + \alpha^{68}x^{55} + \alpha^{119}x^{54} + \alpha^{92}x^{53} + \alpha^{123}x^{52} + \alpha^{177}x^{51} + \alpha^{204}x^{50} + \alpha^{187}x^{49} + \alpha^{254}x^{48} + \alpha^{200}x^{47} + \alpha^{78}x^{46} + \alpha^{141}x^{45} + \alpha^{149}x^{44} + \alpha^{119}x^{43} + \alpha^{26}x^{42} + \alpha^{127}x^{41} + \alpha^{53}x^{40} + \alpha^{160}x^{39} + \alpha^{93}x^{38} + \alpha^{199}x^{37} + \alpha^{212}x^{36} + \alpha^{29}x^{35} + \alpha^{24}x^{34} + \alpha^{145}x^{33} + \alpha^{156}x^{32} + \alpha^{208}x^{31} + \alpha^{150}x^{30} + \alpha^{218}x^{29} + \alpha^{209}x^{28} + \alpha^{4}x^{27} + \alpha^{216}x^{26} + \alpha^{91}x^{25} + \alpha^{47}x^{24} + \alpha^{184}x^{23} + \alpha^{146}x^{22} + \alpha^{47}x^{21} + \alpha^{140}x^{20} + \alpha^{195}x^{19} + \alpha^{195}x^{18} + \alpha^{125}x^{17} + \alpha^{242}x^{16} + \alpha^{238}x^{15} + \alpha^{63}x^{14} + \alpha^{99}x^{13} + \alpha^{108}x^{12} + \alpha^{140}x^{11} + \alpha^{230}x^{10} + \alpha^{242}x^{9} + \alpha^{31}x^{8} + \alpha^{204}x^{7} + \alpha^{11}x^{6} + \alpha^{178}x^{5} + \alpha^{243}x^{4} + \alpha^{217}x^{3} + \alpha^{156}x^{2} + \alpha^{213}x + \alpha^{231}$
66	$x^{66} + \alpha^{5}x^{65} + \alpha^{118}x^{64} + \alpha^{222}x^{63} + \alpha^{180}x^{62} + \alpha^{136}x^{61} + \alpha^{136}x^{60} + \alpha^{162}x^{59} + \alpha^{51}x^{58} + \alpha^{46}x^{57} + \alpha^{117}x^{56} + \alpha^{13}x^{55} + \alpha^{215}x^{54} + \alpha^{81}x^{53} + \alpha^{17}x^{52} + \alpha^{139}x^{51} + \alpha^{247}x^{50} + \alpha^{197}x^{49} + \alpha^{171}x^{48} + \alpha^{95}x^{47} + \alpha^{173}x^{46} + \alpha^{65}x^{45} + \alpha^{137}x^{44} + \alpha^{178}x^{43} + \alpha^{68}x^{42} + \alpha^{111}x^{41} + \alpha^{95}x^{40} + \alpha^{101}x^{39} + \alpha^{41}x^{38} + \alpha^{72}x^{37} + \alpha^{214}x^{36} + \alpha^{169}x^{35} + \alpha^{197}x^{34} + \alpha^{95}x^{33} + \alpha^{7}x^{32} + \alpha^{44}x^{31} + \alpha^{154}x^{30} + \alpha^{77}x^{29} + \alpha^{111}x^{28} + \alpha^{236}x^{27} + \alpha^{40}x^{26} + \alpha^{121}x^{25} + \alpha^{143}x^{24} + \alpha^{63}x^{23} + \alpha^{87}x^{22} + \alpha^{80}x^{21} + \alpha^{253}x^{20} + \alpha^{240}x^{19} + \alpha^{126}x^{18} + \alpha^{217}x^{17} + \alpha^{77}x^{16} + \alpha^{34}x^{15} + \alpha^{232}x^{14} + \alpha^{106}x^{13} + \alpha^{50}x^{12} + \alpha^{168}x^{11} + \alpha^{82}x^{10} + \alpha^{76}x^{9} + \alpha^{146}x^{8} + \alpha^{67}x^{7} + \alpha^{106}x^{6} + \alpha^{171}x^{5} + \alpha^{25}x^{4} + \alpha^{132}x^{3} + \alpha^{93}x^{2} + \alpha^{45}x + \alpha^{105}$
68	$x^{68} + \alpha^{247}x^{67} + \alpha^{159}x^{66} + \alpha^{223}x^{65} + \alpha^{33}x^{64} + \alpha^{224}x^{63} + \alpha^{93}x^{62} + \alpha^{77}x^{61} + \alpha^{70}x^{60} + \alpha^{90}x^{59} + \alpha^{160}x^{58} + \alpha^{32}x^{57} + \alpha^{254}x^{56} + \alpha^{43}x^{55} + \alpha^{150}x^{54} + \alpha^{84}x^{53} + \alpha^{101}x^{52} + \alpha^{190}x^{51} + \alpha^{205}x^{50} + \alpha^{133}x^{49} + \alpha^{52}x^{48} + \alpha^{60}x^{47} + \alpha^{202}x^{46} + \alpha^{165}x^{45} + \alpha^{220}x^{44} + \alpha^{203}x^{43} + \alpha^{151}x^{42} + \alpha^{93}x^{41} + \alpha^{84}x^{40} + \alpha^{15}x^{39} + \alpha^{84}x^{38} + \alpha^{253}x^{37} + \alpha^{173}x^{36} + \alpha^{160}x^{35} + \alpha^{89}x^{34} + \alpha^{227}x^{33} + \alpha^{52}x^{32} + \alpha^{199}x^{31} + \alpha^{97}x^{30} + \alpha^{95}x^{29} + \alpha^{231}x^{28} + \alpha^{52}x^{27} + \alpha^{177}x^{26} + \alpha^{41}x^{25} + \alpha^{125}x^{24} + \alpha^{137}x^{23} + \alpha^{241}x^{22} + \alpha^{166}x^{21} + \alpha^{225}x^{20} + \alpha^{118}x^{19} + \alpha^{2}x^{18} + \alpha^{54}x^{17} + \alpha^{32}x^{16} + \alpha^{82}x^{15} + \alpha^{215}x^{14} + \alpha^{175}x^{13} + \alpha^{198}x^{12} + \alpha^{43}x^{11} + \alpha^{238}x^{10} + \alpha^{235}x^{9} + \alpha^{27}x^{8} + \alpha^{101}x^{7} + \alpha^{184}x^{6} + \alpha^{127}x^{5} + \alpha^{3}x^{4} + \alpha^{5}x^{3} + \alpha^{8}x^{2} + \alpha^{163}x + \alpha^{238}$

Annex B
(normative)

Error correction decoding steps

Take the Version 1-M symbol as an example. For the symbol, the (26, 16, 4) Reed-Solomon code under GF(2^8) is used for error correction. Provided that the code after releasing data masking from the symbol is:

$$R=(r_0,r_1,r_2,\ldots,r_{25})$$

That is,

$$R(x)=r_0+r_1x+r_2x^2+\ \ldots\ +r_{25}x^{25}$$

r_i(i=0-25) is an element of GF(2^8)

(i) Calculate n syndromes (where n is equal to the number of codewords available for error correction, given by ($c - k - p$) as shown in Table 9).

Find the syndrome S_i(i=0-(n − 1)).

$$S_0=R(1)\ =r_0+r_1+r_2+\ \ldots\ +r_{25}$$
$$S_1=R(\alpha)\ =r_0+r_1\alpha+r_2\alpha^2+\ \ldots\ +r_{25}\alpha^{25}$$

...

...

$$S_7=R(\alpha^7)\ =r_0+r_1\alpha^7+r_2\alpha^{14}+\ \ldots\ +r_{25}\alpha^{175}$$

where α is a primitive element of GF(2^8)

(ii) Find the error positions:

$$S_0\sigma_4-S_1\sigma_3+S_2\sigma_2-S_3\sigma_1+S_4=0$$
$$S_1\sigma_4-S_2\sigma_3+S_3\sigma_2-S_4\sigma_1+S_5=0$$
$$S_2\sigma_4-S_3\sigma_3+S_4\sigma_2-S_5\sigma_1+S_6=0$$
$$S_3\sigma_4-S_4\sigma_3+S_5\sigma_2-S_6\sigma_1+S_7=0$$

Find the variable σ_i(i = 1-4) for each error position using the above formulas. Then, substitute the variable for the following polynomial and substitute elements of GF(2^8) one by one.

$$\sigma(x)\ =\ \sigma_4+\ \sigma_3x+\ \sigma_2x^2+\ \sigma_1x^3+x^4$$

Now, it is found that an error is on the jth digit (counting from the 0-th digit) for the element αj which makes σ (α) = 0.

(iii) Find the error size.

Supposing that an error is on the $j1, j2, j4$ digits in (ii) above, then find the size of the error.

$$Y_1\alpha j^1 + Y_2\alpha j^2 + Y_3\alpha j^3 + Y_4\alpha j^4 = S_0$$
$$Y_1\alpha^2 j^1 + Y_2\alpha^2 j^2 + Y_3\alpha^2 j^3 + Y_4\alpha^2 j^4 = S_1$$
$$Y_1\alpha^3 j^1 + Y_2\alpha^3 j^2 + Y_3\alpha^3 j^3 + Y_4\alpha^3 j^4 = S_2$$
$$Y_1\alpha^4 j^1 + Y_2\alpha^4 j^2 + Y_3\alpha^4 j^3 + Y_4\alpha^4 j^4 = S_3$$

Solve the above equations to find the size of each error $Y_i(i = 1\text{-}4)$.

(iv) Correct the error.

Correct the error by adding the complement of the error size value to each error position.

Annex C
(normative)

Format information

C.1 General

The format information consists of a 15-bit sequence comprising 5 data bits and 10 BCH error correction bits. This Annex describes the calculation of the error correction bits and the error correction decoding process.

C.2 Error correction bit calculation

The Bose-Chaudhuri-Hocquenghem (15,5) code shall be used for error correction. The polynomial whose coefficient is the data bit string shall be divided by the generator polynomial $G(x) = x^{10} + x^8 + x^5 + x^4 + x^2 + x + 1$. The coefficient string of the remainder polynomial shall be appended to the data bit string to form the (15,5) BCH code string. Finally, masking shall be applied by XORing the bit string with **101010000010010** (for QR Code symbols) or **100010001000101** (for Micro QR Code symbols) to ensure that the format information bit pattern is not all zeroes for any combination of data mask pattern and Error Correction Level.

EXAMPLE

Error Correction level M; data mask pattern 101

Binary string:	**00101**
Polynomial:	$\mathbf{x^2 + 1}$
Raise power to the (15 - 5) th:	$\mathbf{x^{12} + x^{10}}$
Divide by G(x):	$\mathbf{= (x^{10} + x^8 + x^5 + x^4 + x^2 + x + 1)x^2 + (x^7 + x^6 + x^4 + x^3 + x^2)}$

Add coefficient string of above remainder polynomial to format information data string:

00100 + 0011011100 ′ 001010011011100

XOR with mask	**101010000010010**
Result:	**100000011001110**

Place these bits in the format information areas as described in 7.9.

C.3 Error correction decoding steps

Release the masking of the format information modules by XORing the bit sequence with the mask pattern **101010000010010** (for QR Code symbols) or **100010001000101** (for Micro QR Code symbols).

The Hamming distance of the error correction code used in the format information is 7, which enables up to 3 bit errors to be corrected. There are 32 valid bit sequences for the format information, so decoding by using Table C.1 as a look-up table is efficient. Bit sequences read from the format information area of the symbol are compared with the 32 valid format information bit strings in Table C.1 on a bit by bit basis. The bit string from Table C.1 closest to the bit string read from the symbol is taken, provided the strings differ by 3 bits or less.

Example (for QR Code symbol)

Bit string read from format information area: **000011101001001**

Closest bit string from table: **000111101011001**

Since only 2 bits differ between the two bit strings, the comparison is successful, so the symbol format is confirmed as utilising error correction level M with masking pattern 011.

Table C.1 — Valid format information bit sequences

Sequence before masking		Sequence after masking (QR Code symbols)		Sequence after masking (Micro QR Code symbols)	
Data bits	Error correction bits	binary	hex	binary	hex
00000	0000000000	101010000010010	5412	100010001000101	4445
00001	0100110111	101000100100101	5125	100000101110010	4172
00010	1001101110	101111001111100	5E7C	100111000101011	4E2B
00011	1101011001	101101101001011	5B4B	100101100011100	4B1C
00100	0111101011	100010111111001	45F9	101010110101110	55AE
00101	0011011100	100000011001110	40CE	101000010011001	5099
00110	1110000101	100111110010111	4F97	101111111000000	5FC0
00111	1010110010	100101010100000	4AA0	101101011110111	5AF7
01000	1111010110	111011111000100	77C4	110011110010011	6793
01001	1011100001	111001011110011	72F3	110001010100100	62A4
01010	0110111000	111110110101010	7DAA	110110111111101	6DFD
01011	0010001111	111100010011101	789D	110100011001010	68CA
01100	1000111101	110011000101111	662F	111011001111000	7678
01101	1100001010	110001100011000	6318	111001101001111	734F
01110	0001010011	110110001000001	6C41	111110000010110	7C16
01111	0101100100	110100101110110	6976	111100100100001	7921
10000	1010011011	001011010001001	1689	000011011011110	06DE
10001	1110101100	001001110111110	13BE	000001111101001	03E9
10010	0011110101	001110011100111	1CE7	000110010110000	0CB0
10011	0111000010	001100111010000	19D0	000100110000111	0987
10100	1101110000	000011101100010	0762	001011100110101	1735
10101	1001000111	000001001010101	0255	001001000000010	1202
10110	0100011110	000110100001100	0D0C	001110101011011	1D5B
10111	0000101001	000100000111011	083B	001100001101100	186C
11000	0101001101	011010101011111	355F	010010100001000	2508
11001	0001111010	011000001101000	3068	010000000111111	203F
11010	1100100011	011111100110001	3F31	010111101100110	2F66
11011	1000010100	011101000000110	3A06	010101001010001	2A51
11100	0010100110	010010010110100	24B4	011010011100011	34E3
11101	0110010001	010000110000011	2183	011000111010100	31D4
11110	1011001000	010111011011010	2EDA	011111010001101	3E8D
11111	1111111111	010101111101101	2BED	011101110111010	3BBA

Annex D
(normative)

Version information

D.1 General

The version information consists of an 18-bit sequence comprising 6 data bits and 12 Golay error correction bits. This Annex describes the calculation of the error correction bits and the error correction decoding process.

D.2 Error correction bit calculation

The (18,6) Golay code shall be used for error correction. The polynomial whose coefficient is the data bit string shall be divided by the generator polynomial G(x) = $x^{12} + x^{11} + x^{10} + x^9 + x^8 + x^5 + x^2 + 1$. The coefficient string of the remainder polynomial shall be appended to the data bit string to form the (18,6) Golay code string.

EXAMPLE Version: **7**

Binary string: **000111**

Polynomial: $x^2 + x + 1$

Raise power to the (18 - 6) th: $x^{14} + x^{13} + x^{12}$

Divide by G(x): $= (x^{12} + x^{11} + x^{10} + x^9 + x^8 + x^5 + x^2 + 1)x^2 + (x^{11} + x^{10} + x^7 + x^4 + x^2)$

Add coefficient string of above remainder polynomial to version information data string:

000111 + 110010010100 → 000111110010010100

Place these bits in the version information areas as described in 7.10.

Table D.1 below shows the full version information bit stream for each version.

D.3 Error correction decoding steps

The Hamming distance of the error correction code used in the version information is 8, which enables up to 3 bit errors to be corrected. There are 34 valid bit sequences for the version information, so decoding by using Table D.1 as a look-up table is efficient. Bit sequences read from the version information area of the symbol are compared with the 34 valid version information bit strings in Table D.1 on a bit by bit basis. The bit string from Table D.1 closest to the bit string read from the symbol is taken, provided the strings differ by 3 bits or less after the comparison.

EXAMPLE

Bit string read from version information area: **000111110110010100**

Closest bit string from table: **000111110010010100**

Since only 1 bit differs between the two bit strings, the comparison is successful, so the symbol version is confirmed as 7.

Table D.1 — Version information bit stream for each version

Version	Version information bit stream	Hex equivalent
7	00 0111 1100 1001 0100	07C94
8	00 1000 0101 1011 1100	085BC
9	00 1001 1010 1001 1001	09A99
10	00 1010 0100 1101 0011	0A4D3
11	00 1011 1011 1111 0110	0BBF6
12	00 1100 0111 0110 0010	0C762
13	00 1101 1000 0100 0111	0D847
14	00 1110 0110 0000 1101	0E60D
15	00 1111 1001 0010 1000	0F928
16	01 0000 1011 0111 1000	10B78
17	01 0001 0100 0101 1101	1145D
18	01 0010 1010 0001 0111	12A17
19	01 0011 0101 0011 0010	13532
20	01 0100 1001 1010 0110	149A6
21	01 0101 0110 1000 0011	15683
22	01 0110 1000 1100 1001	168C9
23	01 0111 0111 1110 1100	177EC
24	01 1000 1110 1100 0100	18EC4
25	01 1001 0001 1110 0001	191E1
26	01 1010 1111 1010 1011	1AFAB
27	01 1011 0000 1000 1110	1B08E
28	01 1100 1100 0001 1010	1CC1A
29	01 1101 0011 0011 1111	1D33F
30	01 1110 1101 0111 0101	1ED75
31	01 1111 0010 0101 0000	1F250
32	10 0000 1001 1101 0101	209D5
33	10 0001 0110 1111 0000	216F0
34	10 0010 1000 1011 1010	228BA
35	10 0011 0111 1001 1111	2379F
36	10 0100 1011 0000 1011	24B0B
37	10 0101 0100 0010 1110	2542E
38	10 0110 1010 0110 0100	26A64
39	10 0111 0101 0100 0001	27541
40	10 1000 1100 0110 1001	28C69

Annex E
(normative)

Position of alignment patterns

The alignment patterns are positioned symmetrically on either side of the diagonal running from the top left corner of the symbol to the bottom right corner. They are spaced as evenly as possible between the timing pattern and the opposite side of the symbol, any uneven spacing being accommodated between the timing pattern and the first alignment pattern in the symbol interior.

Table E.1 below shows, for each version, the number of alignment patterns and the row or column coordinates of the center module of each alignment pattern.

Table E.1 — Row/column coordinates of center module of alignment patterns

Version	Number of alignment patterns	Row/Column coordinates of center module					
1	0	-					
2	1	6	18				
3	1	6	22				
4	1	6	26				
5	1	6	30				
6	1	6	34				
7	6	6	22	38			
8	6	6	24	42			
9	6	6	26	46			
10	6	6	28	50			
11	6	6	30	54			
12	6	6	32	58			
13	6	6	34	62			
14	13	6	26	46	66		
15	13	6	26	48	70		
16	13	6	26	50	74		
17	13	6	30	54	78		
18	13	6	30	56	82		
19	13	6	30	58	86		
20	13	6	34	62	90		
21	22	6	28	50	72	94	
22	22	6	26	50	74	98	
23	22	6	30	54	78	102	
24	22	6	28	54	80	106	
25	22	6	32	58	84	110	
26	22	6	30	58	86	114	
27	22	6	34	62	90	118	

Table E.1 *(continued)*

Version	Number of alignment patterns	Row/Column coordinates of center module						
1	0	-						
28	33	6	26	50	74	98	122	
29	33	6	30	54	78	102	126	
30	33	6	26	52	78	104	130	
31	33	6	30	56	82	108	134	
32	33	6	34	60	86	112	138	
33	33	6	30	58	86	114	142	
34	33	6	34	62	90	118	146	
35	46	6	30	54	78	102	126	150
36	46	6	24	50	76	102	128	154
37	46	6	28	54	80	106	132	158
38	46	6	32	58	84	110	136	162
39	46	6	26	54	82	110	138	166
40	46	6	30	58	86	114	142	170

For example, in a Version 7 symbol the table indicates values 6, 22 and 38. The alignment patterns, therefore, are to be centered on (row, column) positions (6,22), (22,6), (22,22), (22,38), (38,22), (38,38). Note that the coordinates (6,6), (6,38), (38,6) are occupied by finder patterns and are not therefore used for alignment patterns.

Annex F
(normative)

Symbology Identifier

The Symbology Identifier assigned to QR Code in ISO/IEC 15424, which should be added as a preamble to the decoded data by a suitably programmed decoder is:

]Qm

where:

] is the Symbology Identifier flag (ASCII value 93)

Q is the code character for the QR Code symbology

m is the modifier character with one of the values defined in Table F.1.

In the case of Micro QR Code symbols, the value of m shall always be 1.

Table F.1 — Symbology Identifier options and modifier values

Modifier value	Option
0	QR Code Model 1 symbol (in accordance with AIM ITS 97-001)
1	QR Code symbol, ECI protocol not implemented
2	QR Code symbol, ECI protocol implemented
3	QR Code symbol, ECI protocol not implemented, FNC1 implied in first position
4	QR Code symbol, ECI protocol implemented, FNC1 implied in first position
5	QR Code symbol, ECI protocol not implemented, FNC1 implied in second position
6	QR Code symbol, ECI protocol implemented, FNC1 implied in second position

The permissible values of m are: 0, 1, 2, 3, 4, 5, 6.

Annex G
(normative)

QR Code print quality – symbology-specific aspects

G.1 General

Because of differences in symbology structures and reference decode algorithms, the effect of certain parameters on a symbol's reading performance may vary. ISO/IEC 15415 provides for symbology specifications to define the grading of certain symbology-specific attributes. This Annex therefore defines the method of grading Fixed Pattern Damage and additional parameters (format information and version information) to be used in the application of ISO/IEC 15415 to QR Code.

G.2 Fixed Pattern damage

G.2.1 Features to be assessed

G.2.1.1 QR Code symbols

The features to be assessed are:

— Three corner segments, each including:

  — the 7 x 7 finder pattern,

  — the 1X wide separators surrounding the two inner sides of the finder pattern,

  — part of the Quiet Zone of a minimum of four modules width (or more if specified by the application) extending for a length of 15 modules along the two outer sides of the finder pattern.

— The two timing patterns of alternating dark and light modules linking the inner corners of the finder patterns.

— The 5 x 5 alignment patterns (where present, in Model 2 symbols of Version 2 or larger).

The features listed above shall be assessed as six segments, viz.:

— the three corner segments (finder patterns with their associated separators and part of the quiet zone) (Segments A1, A2 and A3 respectively),

— the two timing patterns (Segments B1 and B2 respectively),

— the single segment containing all the alignment patterns (Segment C).

Where a timing pattern crosses an alignment pattern the five modules that coincide with the alignment pattern are assessed both as part of the timing pattern and of the alignment pattern.

In a version 7 symbol (45 x 45 modules) for example, each Segment A occupies 168 modules; each Segment B is 29 modules long, and Segment C occupies a total of 150 modules (i.e. 6 x 25).

These segments, in the case of a Version 7 symbol, are illustrated in Figure G.1 below. A1, A2 and A3 indicate the three corner segments; B1 and B2 indicate the two timing pattern segments, and C indicates the single Segment C (comprising the 6 alignment patterns).

NOTE For QR Code symbol its width of Quiet Zone shall be 4X. Figure G.1 shows segments that shall be checked at fixed pattern print quality assessment. Remaining regions of quiet zones are not checked.

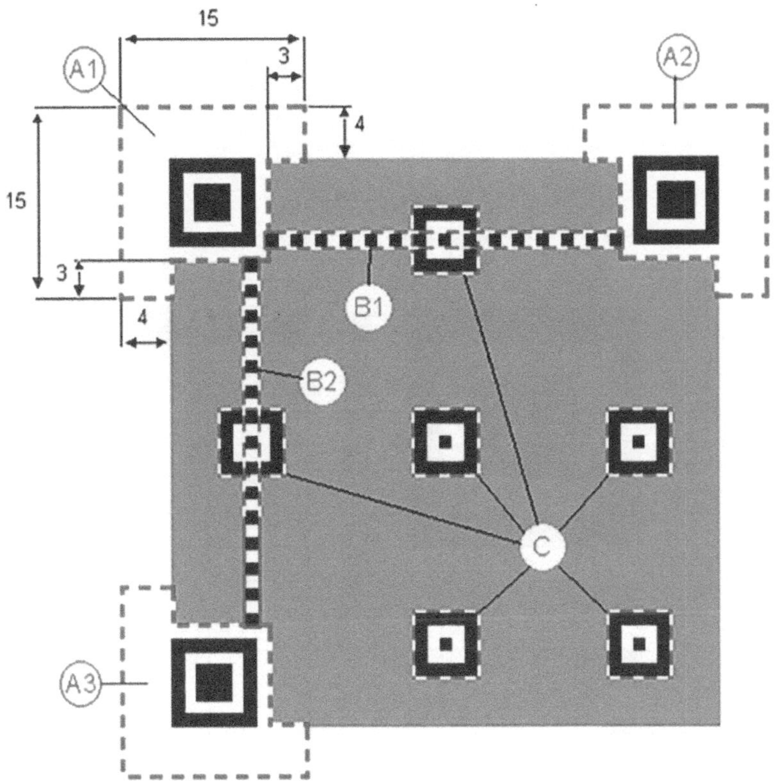

Figure G.1 — QR Code fixed pattern segments

G.2.1.2 Micro QR Code symbols

The features to be assessed are:

- The corner segment, including:
- the finder pattern,
- the 1X wide separators adjoining the two inner sides of the finder pattern,
- part of the Quiet Zone of a minimum of two modules width (or more if specified by the application) extending for a length of 11 modules along the two outer sides of the finder pattern.
- The two timing patterns of alternating dark and light modules running along the top and left side of the symbol from the finder pattern.

The features listed above shall be assessed as three segments, viz.:

- the corner segment (finder pattern with its associated separators and part of the quiet zone) (Segment A), which occupies 104 modules,
- the two timing patterns (Segments B1 and B2 respectively).

In a version M4 symbol (17 x 17 modules) for example, each Segment B is 9 modules long.

These segments, in the case of a Version M4 symbol, are illustrated in Figure G.2 below. A indicates the corner segment; and B1 and B2 indicate the two timing pattern segments.

NOTE For Micro QR Code symbol its width of Quiet Zone shall be 2X. Figure G.2 shows segment that shall be checked at fixed pattern print quality assessment. Remaining regions of quiet zones are not checked.

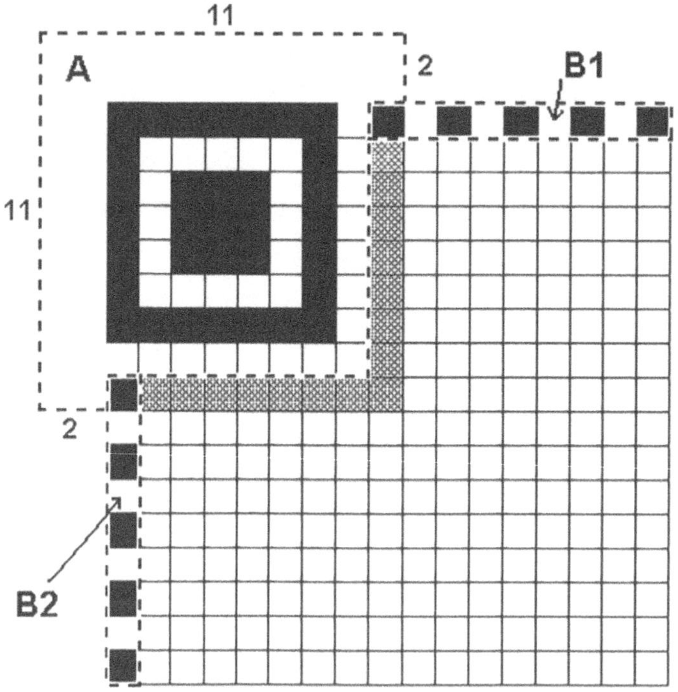

Figure G.2 — Micro QR Code fixed pattern segments

G.2.2 Fixed Pattern Damage grading

Damage to each segment shall be graded based on the modulation of the individual modules that compose it.

The procedure described below shall be applied to each segment in turn

a) From the reference grey-scale image of the symbol, find the modulation grade for each module based on the values in ISO/IEC 15415. Since the intended light or dark nature of the module is known, any module intended to be dark but the reflectance of which is above the global threshold, and any module intended to be light but the reflectance of which is below the global threshold shall be given modulation grade 0.

b) For each modulation grade level, assume that all modules not achieving that grade or a higher grade are module errors, and derive a notional damage grade based on the grade thresholds shown in Table G.1. Take the lower of the modulation grade level and the notional damage grade. The notional damage grade is determined as follows:

   1) For each of Segments A1, A2, and A3, or Segment A in Micro QR Code symbols, count the number of module errors.

2) For segments B1 and B2, count the number of module errors. Express this number as a percentage of the total number of modules in the segment.

3) For segments B1 and B2, taking groups of five adjacent modules and progressing along the segment in steps of one module, verify that in any group of five adjacent modules no more than two are damaged; if this test fails, the grade for the segment shall be 0. This test does not apply to Micro QR Code.

4) For Segment C (in QR Code symbols only), count the number of alignment patterns containing a module error. Express this number as a percentage of the number of alignment patterns in the symbol.

5) Assign a notional damage grade to each segment based on the grade thresholds shown in Table G.1.

c) The Fixed Pattern Damage grade for the segment shall be the highest resulting grade for all modulation grade levels.

The Fixed Pattern Damage grade for the symbol shall be the lowest of the segment grades.

Table G.1 — Grade thresholds for QR Code Fixed Pattern Damage

Segments A1, A2 and A3 (QR Code); Segment A (Micro QR Code)	Segments B1 and B2 (QR Code)	Segments B1 and B2 (Micro QR Code)	Segment C (QR Code)	Grade
Number of module errors	Percentage of total modules with module errors	Percentage of total modules with module errors	Percentage of alignment patterns with module errors	
0	0%	0%	0%	4
1	≤ 7%		≤ 10%	3
2	≤ 11%	≤30%	≤ 20%	2
3	≤ 14%		≤ 30%	1
≥ 4	> 14%	>30%	> 30%	0

G.3 Grading of additional parameters

G.3.1 General

QR Code symbols contain a duplicated set of modules representing information that defines the format of the symbol, and symbols of Version 7 to 40 also contain a duplicated set of modules representing information that defines the symbol size. Micro QR Code symbols contain a single set of modules representing information that defines the format of the symbol. This data requires to be reliably detected at an early stage of the decoding procedure, and if it cannot be decoded, the remainder of the symbol cannot be decoded. For this reason the format information and version information module blocks are graded separately (in a similar way to Fixed Pattern Damage), and their grades are included in the overall symbol grade determination.

G.3.2 Grading of format information

For each block of format information, determine a grade for the block according to the following method.

a) From the reference grey-scale image of the symbol, find the modulation grade for each module based on the values in ISO/IEC 15415. Since the intended light or dark nature of the module is known after decode, any module intended to be dark but the reflectance of which is above the global threshold, and any module intended to be light but the reflectance of which is below the global threshold shall be given modulation grade 0. If the format information in the block cannot be decoded, the grade for the block shall be 0.

b) For each modulation grade level:

1) Assume that all modules not achieving that modulation grade or a higher grade are module errors, and derive a notional grade based on Table G.2:

Table G.2 — Format information notional grading

Number of module errors	Grade
0	4
1	3
2	2
3	1
≥ 4	0

2) Select the lower of the MOD grade and the notional grade at each level as the grade for that level, as illustrated in Table G.3.

3) The grade for the block shall be the highest resulting grade, as illustrated in Table G.3.

Table G.3 — Example of grading of format information block

Modulation grade	Notional grade	Lower of grades
4	2	2
3	2	2
2	3	2
1	3	1
0	4	0
	Selected (highest) Grade->	2

c) The format information grade shall be:

1) For QR Code symbols, the average of the grades of the two format information blocks, rounded up if necessary to the next integer.

2) For Micro QR Code symbols, the grade determined in step 2 c).

G.3.3 Grading of version information (QR Code symbols)

For each block of version information, determine a grade for the block according to the following method.

a) Find the modulation grade for each module based on the values in ISO/IEC 15415. Since the intended light or dark nature of the module is known after decode, any module intended to be dark but the reflectance of which is above the global threshold, and any module intended to be light but the reflectance of which is below the global threshold shall be given modulation grade 0. If the version information in the block cannot be decoded, the grade for the block shall be 0.

b) For each modulation grade level:

1) Assume that all modules not achieving that modulation grade or a higher grade are module errors, and derive a notional grade based on Table G.4:

Table G.4 — Version information notional grading

Number of module errors	Grade
0	4
1	3
2	2
3	1
≥4	0

2) Select the lower of the MOD grade and the notional grade at each level as the grade for that level, as illustrated in Table G.5.

3) The grade for the block shall be the highest resulting grade, as illustrated in Table G.5.

Table G.5 — Example of grading of version information block

Modulation grade	Notional grade	Lower of grades
4	2	2
3	2	2
2	3	2
1	3	1
0	4	0
	Selected (highest) Grade->	2

c) The version information grade shall be the average of the grades of the two version information blocks, rounded up if necessary to the next integer.

G.4 Scan grade

The scan grade shall be the lowest of the grades for the standard parameters evaluated according to ISO/IEC 15415 together with the grades for Fixed Pattern Damage, format information and (where applicable) version information evaluated in accordance with this Annex.

Annex H
(informative)

JIS8 and Shift JIS character sets

Table H.1 — 8-bit character set for JIS X 0201 (JIS8)

Char.	Hex	Char.	Hex	Char.	Hex	Char.	Hex	Char.	Hex	Char.	Hex	Char.	Hex	Char.	Hex
NUL	00	SP	20	@	40	`	60		80		A0	タ	C0		E0
SOH	01	!	21	A	41	a	61		81	。	A1	チ	C1		E1
STX	02	"	22	B	42	b	62		82	「	A2	ツ	C2		E2
ETX	03	#	23	C	43	c	63		83	」	A3	テ	C3		E3
EOT	04	$	24	D	44	d	64		84	、	A4	ト	C4		E4
ENQ	05	%	25	E	45	e	65		85	・	A5	ナ	C5		E5
ACK	06	&	26	F	46	f	66		86	ヲ	A6	ニ	C6		E6
BEL	07	'	27	G	47	g	67		87	ァ	A7	ヌ	C7		E7
BS	08	(	28	H	48	h	68		88	ィ	A8	ネ	C8		E8
HT	09	)	29	I	49	I	69		89	ゥ	A9	ノ	C9		E9
LF	0A	*	2A	J	4A	j	6A		8A	ェ	AA	ハ	CA		EA
VT	0B	+	2B	K	4B	k	6B		8B	ォ	AB	ヒ	CB		EB
FF	0C	,	2C	L	4C	l	6C		8C	ャ	AC	フ	CC		EC
CR	0D	-	2D	M	4D	m	6D		8D	ュ	AD	ヘ	CD		ED
SO	0E	.	2E	N	4E	n	6E		8E	ョ	AE	ホ	CE		EE
SI	0F	/	2F	O	4F	o	6F		8F	ッ	AF	マ	CF		EF
DLE	10	0	30	P	50	p	70		90	ー	B0	ミ	D0		F0
DC1	11	1	31	Q	51	q	71		91	ア	B1	ム	D1		F1
DC2	12	2	32	R	52	r	72		92	イ	B2	メ	D2		F2
DC3	13	3	33	S	53	s	73		93	ウ	B3	モ	D3		F3
DC4	14	4	34	T	54	t	74		94	エ	B4	ヤ	D4		F4
NAK	15	5	35	U	55	u	75		95	オ	B5	ユ	D5		F5
SYN	16	6	36	V	56	v	76		96	カ	B6	ヨ	D6		F6
ETB	17	7	37	W	57	w	77		97	キ	B7	ラ	D7		F7
CAN	18	8	38	X	58	x	78		98	ク	B8	リ	D8		F8
EM	19	9	39	Y	59	y	79		99	ケ	B9	ル	D9		F9
SUB	1A	:	3A	Z	5A	z	7A		9A	コ	BA	レ	DA		FA
ESC	1B	;	3B	[	5B	{	7B		9B	サ	BB	ロ	DB		FB
FS	1C	<	3C	¥	5C	|	7C		9C	シ	BC	ワ	DC		FC
GS	1D	=	3D	]	5D	}	7D		9D	ス	BD	ン	DD		FD
RS	1E	>	3E	^	5E	‾	7E		9E	セ	BE	゛	DE		FE
US	1F	?	3F	_	5F	DEL	7F		9F	ソ	BF	゜	DF		FF

Figure H.1 below shows the areas of the 256 x 256 code plane occupied by Shift JIS double byte characters.

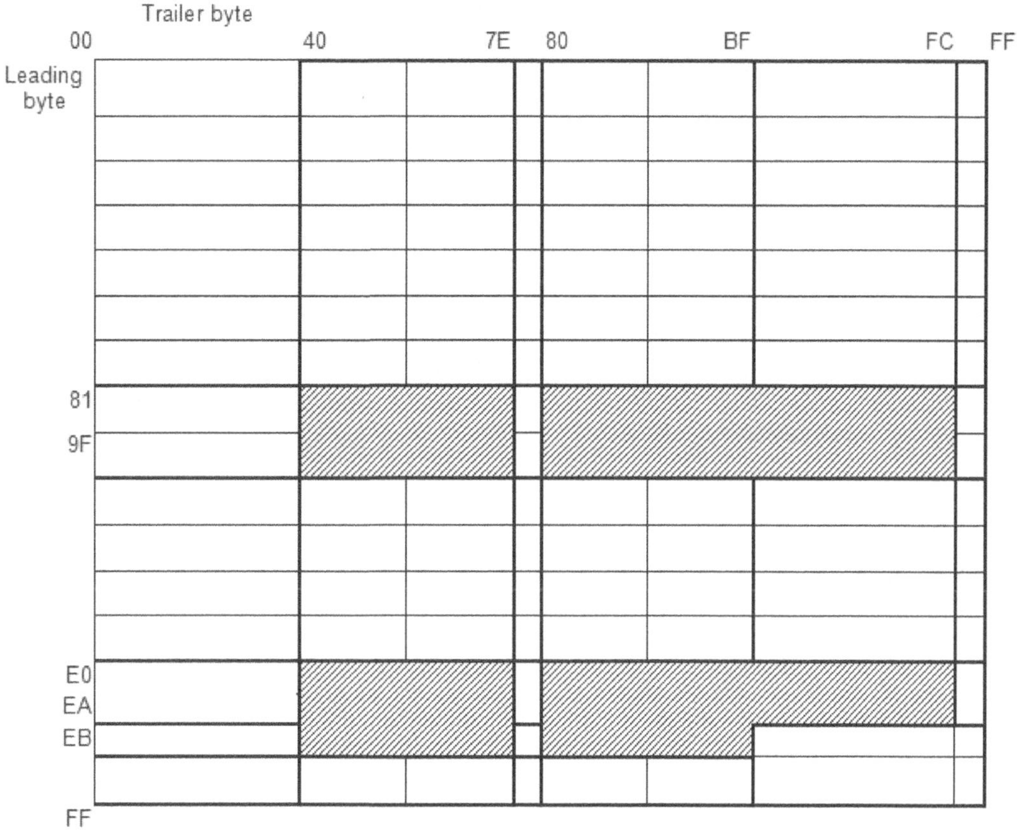

Figure H.1 — Shift JIS character values

According to JIS X 0208:1997, Annex 1, leading and trailing bytes within the ranges shown shaded are assigned to Shift JIS Kanji characters. Any pairs of bytes within these ranges may be encoded using the Kanji mode compaction scheme.

Annex I
(informative)

Symbol encoding examples

I.1 General

This Annex describes the encoding of the data string **01234567** into both a QR Code symbol and a Micro QR Code symbol.

I.2 Encoding a QR Code symbol

The data string is to be encoded into a version 1-M symbol, using the Numeric mode in accordance with 7.4.3.

Step 1: Data Encoding

- Divide into groups of three digits and convert each group to its 10 or 7-bit binary equivalent:
  - 012 → 0000001100
  - 345 → 0101011001
  - 67 → 1000011
- Convert character count indicator to binary (10 bits for version 1-M)

Character count indicator (8) = **0000001000**

- Connect mode indicator for Numeric mode (**0001**), character count indicator, binary data, and Terminator (**0000**)

0001 0000001000 0000001100 0101011001 1000011 0000

- Divide into 8-bit codewords, adding padding bits (shown underlined for illustration) as needed

00010000 00100000 00001100 01010110 01100001 10000000

- Add Pad codewords to fill data codeword capacity of symbol (for version 1-M, 16 data codewords, therefore 10 Pad codewords required (shown underlined for illustration)), giving the result:

00010000 00100000 00001100 01010110 01100001 10000000 11101100 00010001 11101100 00010001 11101100 00010001 11101100 00010001 11101100 00010001

Step 2: Error Correction Codeword generation

Using the Reed-Solomon algorithm to generate the required number of error correction codewords (for a Version 1-M symbol, 10 are needed), these (shown underlined for illustration) should be added to the bit stream, resulting in:

00010000 00100000 00001100 01010110 01100001 10000000 11101100 00010001 11101100 00010001 11101100 00010001 11101100 00010001 11101100 00010001 10100101 00100100 11010100 11000001 11101101 00110110 11000111 10000111 00101100 01010101

Step 3: Module placement in matrix

As there is only a single error correction block in a version 1-M symbol, no interleaving is required in this instance. The finder patterns, separators and timing patterns are placed in a blank 21 × 21 matrix and the module positions for the format information are left temporarily blank. The codewords from Step 2 are placed in the matrix in accordance with 7.7.3, which results in the arrangement shown in Figure I.1.

Figure I.1 — Data modules placed in symbol prior to data masking

Step 4: Data masking pattern selection

Apply the data masking patterns defined in 7.8.2 in turn and evaluate the results in accordance with 7.8.3. The data masking pattern selected is referenced **010**.

Step 5: Format information

The error correction level is M and the data masking pattern is 011. Therefore, from 7.9.1 the data bits of the format information are **00 010**.

The BCH error correction calculation gives **1001101110** as the bit sequence to be added to the data, giving:

000101001101110 as the unmasked format information.

XOR this bit stream with the mask **101010000010010**:

000101001101110 (raw bit stream)

101010000010010 (mask)

101111001111100 (format information to be placed in symbol)

Step 6: Final symbol construction

Apply the selected data masking pattern to the encoding region of the symbol as described in 7.8, and add format information modules in positions reserved in step 3. The final symbol is shown in Figure I.2.

Figure I.2 — Final version 1-M symbol encoding 01234567

I.3 Encoding a Micro QR Code symbol

The data string **01234567** is to be encoded into a Version M2 symbol with EC level L, using the Numeric mode in accordance with 7.4.3.

Step 1: Data Encoding

— Divide into groups of three digits and convert each group to its 10 or 7-bit binary equivalent:

— 012 → 0000001100

— 345 → 0101011001

— 67 → 1000011

— Mode indicator for Numeric mode in Version M2 is **0**

— Character count is 8; convert to binary (4 bits for Version M2-L):

Character count indicator (8) = **1000**

— Terminator for Version M2 is 5 zero bits, **00000**

— Connect mode indicator for Numeric mode (**0**), character count indicator (**1000**), binary data, and Terminator (**00000**)

0 1000 0000001100 0101011001 1000011 00000

— Divide into 8-bit codewords, adding 3 padding bits (shown underlined for illustration) since final codeword contained only 5 bits

01000000 00011000 10101100 11000011 00000<u>000</u>

— No Pad codewords are required to fill data codeword capacity of symbol (for version M2-L, 5 data codewords).

Step 2: Error Correction Codeword generation

Using the Reed-Solomon algorithm to generate the required number of error correction codewords (for a Version M2-L symbol, 5 are needed), these (shown underlined for illustration) should be added to the bit stream, resulting in:

01000000 00011000 10101100 11000011 00000000 10000110 00001101 00100010 10101110 00110000

Step 3: Module placement in matrix

The finder pattern and timing patterns are placed in a blank 13 × 13 matrix and the module positions for the format information are left temporarily blank. The codewords from Step 2 are placed in the matrix in accordance with 7.7.3. Figure I.3 shows the module arrangement.

Figure I.3 — Data modules placed in symbol prior to data masking

Step 4: Data masking pattern selection

Apply the data masking patterns defined in 7.8.2 in turn and evaluate the results in accordance with 7.8.3. The data masking pattern selected is referenced **01**. Apply the selected data masking pattern to the encoding region of the matrix, as described in 7.8.

Step 5: Format information

The symbol number for an M2-L symbol is 1, which is represented in binary form as **001**, and the data masking pattern is **01**. Therefore, the data bits of the format information are **001 01**.

The BCH error correction calculation gives **0011011100** as the bit sequence to be added to the data, giving:

001010011011100 as the unmasked format information.

XOR this bit stream with the mask **100010001000101**:

— **001010011011100** (raw bit stream)

— **100010001000101** (mask)

— **101000010011001** (format information to be placed in symbol)

Step 6: Final symbol construction

Add format information modules in positions reserved in step 3. The final symbol is shown in Figure I.4.

Figure I.4 — Final version M2-L symbol encoding 01234567

Annex J
(informative)

Optimisation of bit stream length

J.1 General

As described in this standard, QR Code offers various modes of encoding each of which differs in the number of bits it requires to represent a given data string. Since there is an overlap between the character sets of each mode - for example, numeric data may be encoded in Numeric, Alphanumeric and Byte modes, and Latin alphanumeric data may be encoded in Alphanumeric and Byte modes - the symbol generation software may need to choose the most appropriate mode in which to encode data characters which appear in more than one mode.

A choice may also be possible between a QR Code symbol and a Micro QR Code symbol.

The choice of mode must be made initially and the mode may be changed part way through a data stream.

A number of alternative approaches may be adapted to minimize the bit stream length. The algorithm will need not only to consider the immediate sequence of characters but also look ahead to the next sequence of data in view of the overhead required for switching modes. The term "exclusive subset" is used in this Annex as a short way of referring to the set of characters within the character set of a mode which are not shared with the more restricted character set of another mode, as shown below and in Table J.1.

The numeric exclusive subset is the set of hex values 30 to 39 (digits 0 to 9).

The Alphanumeric exclusive subset is the set of hex values 20, 24, 25, 2A, 2B, 2D to 2F, 3A, and 41 to 5A, mapped as {A - Z, space, $ % * + - . / :}.

NOTE 1 This subset does not include the digits.

The Byte exclusive subset comprises hex values 00 - FF, but excludes hex values 20, 24, 25, 2A, 2B, 2D - 3A, and 41 - 5A.

NOTE 2 The excluded values are contained in the alphanumeric and numeric exclusive subsets.

Table J.1 — Exclusive subset byte values for QR Code modes

Exclusive subset	Byte values (hex)
Numeric	30 to 39
Alphanumeric	20, 24, 25, 2A, 2B, 2D to 2F, 3A, and 41 to 5A
Byte	00 to 1F, 21 to 23, 26 to 29, 2C, 3B to 40, 5B to FF (excluding reserved values 80 to 9F and E0 to FF)
Kanji	All double bytes in ranges defined in Annex H

The compaction efficiencies given in 7.4.3 to 7.4.6 need to be interpreted carefully. The best scheme for a given set of data may not be the one with the fewest bits per data character. If the highest degree of compaction is required, account has to be taken of the additional bits required to change modes (additional mode indicator and character count indicator). It should also be noted that even if the number of codewords is minimized, the codeword stream may need to be expanded to fill a symbol. This fill process is done using pad characters.

J.2 Optimisation for QR Code symbols

For QR Code symbols, the following guidelines form the basis of one possible algorithm to determine the shortest bit stream for any given input data.

Numbers of characters shown in square brackets e.g.[5,7,9] are applicable to versions 1 - 9, 10 - 26, and 27 - 40 respectively.

a) Select initial mode:

   1) If initial input data is in the exclusive subset of the Byte character set, select Byte mode;

   2) If initial input byte is in the Kanji leading byte exclusive subset and the next byte is in the Kanji trailing byte exclusive subset, AND the subsequent data is in the Alphanumeric or Numeric exclusive character set, select Kanji mode, ELSE if subsequent data is in the Byte exclusive character set AND the following[5,5,6] byte pairs are also in the Kanji exclusive subsets, select Byte mode;

   3) If initial input data is in the exclusive subset of the Alphanumeric character set AND if there are less than[6-8] characters followed by data from the remainder of the Byte character set, THEN select the Byte mode ELSE select Alphanumeric mode;

   4) If initial data is numeric, AND if there are less than[4,4,5] characters followed by data from the exclusive subset of the Byte character set, THEN select Byte mode ELSE IF there are less than[7-9] characters followed by data from the exclusive subset of the Alphanumeric character set THEN select Alphanumeric mode ELSE select Numeric mode.

b) While in Byte mode:

   1) If a sequence of at least[9,12,13] byte pairs from the Kanji set occurs before more data from the exclusive subset of the Byte character set, switch to Kanji mode;

   2) If a sequence of at least[11,15,16] character from the exclusive subset of the Alphanumeric character set occurs before more data from the exclusive subset of the Byte character set, switch to Alphanumeric mode;

   3) If a sequence of at least[6,8,9] Numeric characters occurs before more data from the exclusive subset of the Byte character set, switch to Numeric mode;

   4) If a sequence of at least[6-8] Numeric characters occurs before more data from the exclusive subset of the Alphanumeric character set, switch to Numeric mode.

c) While in Alphanumeric mode:

   1) If one or more Kanji characters occurs, switch to Kanji mode;

   2) If one or more characters from the exclusive subset of the Byte character set occurs, switch to Byte mode;

   3) If a sequence of at least[13,15,17] Numeric characters occurs before more data from the exclusive subset of the Alphanumeric character set, switch to Numeric mode.

d) While in Numeric mode:

   1) If one or more Kanji character occurs, switch to Kanji mode;

   2) If one or more characters from the exclusive subset of the Byte character set occurs, switch to Byte mode;

   3) If one or more characters from the exclusive subset of the Alphanumeric character set occurs, switch to Alphanumeric mode.

J.3 Optimisation for Micro QR Code symbols

J.3.1 Optimisation principles

Assuming that the data to be encoded is in the exclusive subsets of not more than two modes, and that all the data in each subset is grouped together (e.g. "123abcdef"), an algorithm to determine the shortest bit stream for Micro QR Code data can be derived from Table J.2. These principles can be extended to cater for more than two modes, although care must be taken that the resulting bit stream will fit one of the available symbols.

Because the lower modes use fewer bits per character than the higher modes, there is a point at which the extra overhead of the additional mode indicator and character count indicator for a change of mode is offset by the greater encoding density of the lower mode. Table J.2 shows the minimum number of consecutive characters in a lower mode for which a shorter total bit stream is achieved by changing modes. For fewer characters, encoding all the data in the higher mode will give a shorter bit stream.

Table J.2 — Minimum characters in lower mode for minimising bit stream length by changing modes

Mode combination	M2 symbols	M3 symbols	M4 symbols
Numeric + Alphanumeric	3 numeric	4 numeric	5 numeric
Numeric + 8-bit byte	n/a	2 numeric	3 numeric
Alphanumeric + Byte	n/a	3 alphanumeric	4 alphanumeric

J.3.2 Capacity of Micro QR Code symbols

Based on the principles of the above table, and the capacities of the various symbol versions, Figures J.1 to J.6 below show, for each combination of modes, the options available for encoding given amounts of data in combinations of modes.

The column and row headings identify the number of characters in each mode. The figures show the symbol versions and error correction levels, omitting the initial M; thus, for example, 4Q refers to a version M4 symbol with error correction level Q. For any given combination of characters and modes, the available symbol versions are those at the appropriate row and column intersection and those shown to the right of or below that intersection.

For example, if the data string was "123456ABCDEFGH", consisting of six numeric characters and eight from the alphanumeric character set, Figure J.1 shows that the data would fit into a version M3-L symbol (total of 77 bits including mode indicators and character count indicators), or a version M4-M symbol or a version M4-L symbol (81 bits for either). The options may be narrowed down either by the space available or the required level of error correction.

	Alphanumeric																					
Num.	0	1	2	3	4	5	6	7	8	9	10	11	12	13	14	15	16	17	18	19	20	21
0						2M	2L					3M		4Q	3L				4M			4L
1	1				2M	2L					3M		4Q	3L				4M			4L	
2	1			2M	2L					3M		4Q	3L				4M			4L		
3	1		2M	2L					3M		4Q	3L				4M			4L			
4	1	2M		2L				3M		4Q	3L				4M			4L				
5	1	2M	2L				3M		4Q	3L					4M			4L				
6	2M		2L				3M	4Q		3L				4M			4L					
7	2M	2L				3M		4Q	3L				4M			4L						
8	2M					3M	4Q		3L				4M			4L						
9	2L				3M		4Q	3L				4M			4L							
10	2L			3M		4Q	3L					4M			4L							
11				3M	4Q	3L					4M			4L								
12			3M		4Q	3L					4M		4L									
13			3M	4Q	3L					4M			4L									
14		3M	4Q		3L				4M			4L										
15	3M		4Q	3L					4M			4L										
16	3M	4Q		3L				4M			4L											
17	3M	4Q	3L				4M			4L												
18	3M/4Q		3L				4M			4L												
19	4Q	3L				4M			4L													
20	3L/4Q					4M			4L													
21	3L/4Q				4M			4L														
22	3L			4M			4L															
23	3L			4M			4L															
24			4M			4L																
25			4M		4L																	
26		4M			4L																	
27	4M			4L																		
28	4M			4L																		
29	4M		4L																			
30	4M		4L																			
31		4L																				
32	4L																					
33	4L																					
34	4L																					
35	4L																					

Figure J.1 — Micro QR Code symbol capacities - numeric and alphanumeric data

Num.	Byte															
	0	1	2	3	4	5	6	7	8	9	10	11	12	13	14	15
0								3M		3L/4Q				4M		4L
1							3M		3L/4Q				4M		4L	
2							3M	4Q	3L			4M		4L		
3						3M	4Q	3L			4M		4L			
4						3M	4Q	3L			4M		4L			
5	1				3M	4Q	3L			4M		4L				
6	2M				3M	4Q	3L			4M		4L				
7	2M			3M	4Q	3L			4M		4L					
8	2M			3M	4Q	3L			4M		4L					
9	2L			3M	4Q	3L			4M		4L					
10	2L		3M	4Q	3L			4M		4L						
11			3M	4Q	3L			4M		4L						
12		3M	4Q	3L			4M		4L							
13		3M	4Q	3L			4M		4L							
14		3M	4Q	3L			4M		4L							
15	3M	4Q	3L			4M		4L								
16	3M	4Q	3L			4M		4L								
17	3M/4Q	3L			4M		4L									
18	3M/4Q	3L			4M		4L									
19	3L/4Q			4M		4L										
20	3L/4Q			4M		4L										
21	3L/4Q			4M		4L										
22	3L		4M		4L											
23	3L		4M		4L											
24		4M		4L												
25		4M		4L												
26		4M		4L												
27	4M		4L													
28	4M		4L													
29	4M	4L														
30	4M	4L														
31	4L															
32	4L															
33	4L															
34	4L															
35	4L															

Figure J.2 — Micro QR Code symbol capacities - numeric and Byte data

A/num.	Byte															
	0	1	2	3	4	5	6	7	8	9	10	11	12	13	14	15
0								3M		3L/4Q				4M		4L
1	2M						3M		3L/4Q				4M		4L	
2	2M					3M		3L/4Q				4M		4L		
3	2M				3M		3L/4Q				4M		4L			
4	2M				3M	4Q	3L			4M		4L				
5	2M			3M	4Q	3L			4M		4L					
6	2L		3M	4Q	3L			4M		4L						
7			3M	4Q	3L			4M		4L						
8		3M	4Q	3L			4M		4L							
9	3M	4Q	3L			4M		4L								
10	3M	4Q	3L			4M		4L								
11	3M/4Q	3L			4M		4L									
12	3L/4Q			4M		4L										
13	3L/4Q			4M		4L										
14	3L		4M		4L											
15		4M		4L												
16		4M		4L												
17	4M		4L													
18	4M	4L														
19	4L															
20	4L															
21	4L															

Figure J.3 — Micro QR Code symbol capacities - alphanumeric and Byte data

Num.	Kanji 0	1	2	3	4	5	6	7	8	9
0					3M	4Q	3L		4M	4L
1	1				3M/4Q	3L		4M	4L	
2	1			3M	4Q	3L	4M		4L	
3	1			3M	3L/4Q		4M	4L		
4	1			3M/4Q	3L		4M	4L		
5	1			3M/4Q	3L		4M	4L		
6	2M		3M	4Q	3L	4M		4L		
7	2M		3M	3L/4Q		4M	4L			
8	2M		3M/4Q	3L		4M	4L			
9	2L		3M/4Q	3L		4M	4L			
10	2L	3M	3L/4Q		4M		4L			
11		3M	3L/4Q		4M	4L				
12		3M/4Q	3L		4M	4L				
13	3M	4Q	3L		4M	4L				
14	3M	3L/4Q		4M		4L				
15	3M	3L/4Q		4M	4L					
16	3M/4Q	3L		4M	4L					
17	3M/4Q	3L		4M	4L					
18	3M/4Q		4M		4L					
19	3L/4Q		4M	4L						
20	3L/4Q		4M	4L						
21	3L/4Q		4M	4L						
22	3L	4M	4L							
23	3L	4M	4L							
24		4M	4L							
25	4M		4L							
26	4M	4L								
27	4M	4L								
28	4M	4L								
29	4M	4L								
30	4M									
31	4L									
32	4L									
33	4L									
34	4L									
35	4L									

Figure J.4 — Micro QR Code symbol capacities - numeric and Kanji data

	Kanji									
A/num.	0	1	2	3	4	5	6	7	8	9
0					3M	4Q	3L		4M	4L
1	2M			3M	4Q	3L		4M	4L	
2	2M			3M	3L/4Q		4M	4L		
3	2M			3M/4Q	3L		4M	4L		
4	2M		3M	3L/4Q		4M		4L		
5	2M		3M/4Q	3L		4M	4L			
6	2L	3M	4Q	3L	4M		4L			
7		3M	3L/4Q		4M	4L				
8		3M/4Q	3L		4M	4L				
9	3M	3L/4Q		4M	4L					
10	3M/4Q	3L		4M	4L					
11	3M/4Q		4M		4L					
12	3L/4Q		4M	4L						
13	3L/4Q	4M		4L						
14	3L	4M	4L							
15		4M	4L							
16	4M	4L								
17	4M	4L								
18	4M									
19	4L									
20	4L									
21	4L									

Figure J.5 — Micro QR Code symbol capacities - alphanumeric and Kanji data

	Kanji									
8-bit	0	1	2	3	4	5	6	7	8	9
0					3M	4Q	3L		4M	4L
1				3M	4Q	3L		4M	4L	
2				3M/4Q	3L		4M	4L		
3			3M	3L/4Q		4M	4L			
4		3M	4Q	3L		4M	4L			
5		3M/4Q	3L		4M	4L				
6	3M	3L/4Q		4M		4L				
7	4Q	3L		4M	4L					
8	3L/4Q		4M	4L						
9	3L/4Q	4M		4L						
10		4M	4L							
11	4M	4L								
12	4M	4L								
13	4M									
14	4L									
15	4L									

Figure J.6 — Micro QR Code symbol capacities - Byte and Kanji data

Annex K
(informative)

User guidelines for printing and scanning of QR Code symbols

K.1 General

Any QR Code application must be viewed as a total system solution. All the symbology encoding/decoding components (surface markers or printers, labels, readers) making up an installation need to operate together as a system. A failure in any link of the chain, or a mismatch between them, could compromise the performance of the overall system.

While compliance with the specifications is one key to assuring overall system success, other considerations come into play which may influence performance as well. The following guidelines suggest some factors to keep in mind when specifying or implementing bar or matrix code systems:

a) Select a print density which will yield tolerance values that can be achieved by the marking or printing technology being used. Ensure that the module dimension is an integer multiple of the print head pixel dimension (both parallel to and perpendicular to the print direction). Ensure also that any adjustment for print gain (or loss) is performed by changing an equal integer number of pixels from dark to light (or light to dark) on all dark-to-light boundaries of individual or groups of adjoining dark modules in order to ensure that the module center spacing remains constant, although the apparent bit-map representation of the individual dark (or light) modules is adjusted in size to suit the direction of compensation.

b) Choose a reader with a resolution suitable for the symbol density and quality produced by the marking or printing technology.

c) Ensure that the optical properties of the printed symbol are compatible with the wavelength of the scanner light source or sensor.

d) Verify symbol compliance in the final label or package configuration. Overlays, show-through and curved or irregular surfaces can all affect symbol readability.

The effects of specular reflection from glossy symbol surfaces must be considered. Scanning systems must take into account the variations in diffuse reflection between dark and light features. At some scanning angles, the specular component of the reflected light can greatly exceed the desired diffuse component, changing the scanning performance. In cases where the surface of the material or part can be altered, matt, non-glossy surfaces may help minimize specular effects. Where this option is not available, particular must be taken to ensure the illumination of the symbol to be read optimizes the desired contrast components.

K.2 User selection of error correction level

The users should define the appropriate level of error correction to suit the application requirements. As shown in Table 8, the four levels from L to H offer increasing capabilities of detecting and correcting errors, at the cost of some increase in symbol size for a given message length. For example, a Version 20-Q symbol can contain a total of 485 data codewords, but if a lower level of error correction was acceptable, the same data could also be represented in a Version 15-L symbol (exact capacity 523 data codewords).

The error correction level should be determined in relation to:

— the expected level of symbol quality: the lower the expected quality grade, the higher the level to be applied;

- the importance of a high first read rate;
- the opportunity for re-scanning in the event of a read failure;
- the space constraints which might reduce the opportunity to use a higher error correction level.

Error correction level L is appropriate for high symbol quality and/or the need for the smallest possible symbol for given data. Level M is described as "Standard" level and offers a good compromise between small size and increased reliability. Level Q is a "High reliability" level and suitable for more critical or poor print quality applications while level H offers the maximum achievable reliability.

Annex L
(informative)

Autodiscrimination

QR Code may be read by suitably programmed decoders which have been designed to autodiscriminate it from other symbologies. A properly programmed QR Code reader will not decode a symbol in another symbology as a valid QR Code symbol; however, representations of short linear symbols may be found in any matrix symbol including QR Code.

Although QR Code Model 1 symbols can be autodiscriminated from QR Code symbols by a suitable decoder, it is strongly recommended that the two types of symbol should not be mixed in an application.

The decoder's valid set of symbologies should be limited to those needed by a given application in order to maximize reading security.

Annex M
(informative)

Process control techniques

M.1 General

This Annex describes tools and procedures useful for monitoring and controlling the process of creating scannable QR Code symbols. These techniques do not constitute a print quality check of the produced symbols - the method defined in 10 and Annex G is the required method for assessing symbol quality - but they individually and collectively yield good indications of whether the symbol production process is creating workable symbols.

M.2 Symbol Contrast

Most verifiers for linear bar code symbols have either a reflectometer mode or a mode for plotting scan reflectance profiles and/or reporting Symbol Contrast, as defined in ISO/IEC 15416, from undecodable scans. Except with symbols requiring special illumination configurations, the symbol contrast readings that can be obtained using a 0,150 mm or 0,250 mm aperture at 660 nm wavelength - either the reported symbol contrast value, the maximum to minimum scan reflectance profile excursions, or the difference between maximum and minimum reflectometer readings - are found to correlate well with an image-derived symbol contrast value. In particular these reading can be used to check that symbol contrast stays well above the minimum allowed for the intended symbol quality grade.

M.3 Assessing Axial Nonuniformity

For a QR Code symbol, measure the distance from the left edge of the upper left finder pattern to the right edge of the upper right finder pattern, and the distance from the top edge of the upper left finder pattern to the bottom edge of the lower left finder pattern. For a Micro QR Code symbol, measure the distance from the left edge of the upper left finder pattern to the right edge of the rightmost module in the upper timing pattern, and the distance from the top edge of the upper left finder pattern to the bottom edge of the lowest module in the left side timing pattern. Divide each of these by the number of modules in that dimension. E.g. a version 2 symbol would have 25 as a divisor. Substitute the results for X_{AVG} and Y_{AVG} in the formula in G.2.4 and grade the result for an assessment of Axial Nonuniformity.

M.4 Visual inspection for symbol distortion and defects

Ongoing visual inspection of the Finder and timing patterns in sample symbols can monitor an important aspect of the production process.

Matrix code symbols are susceptible to errors caused by local distortions of the matrix grid. Any such distortions may show up visually as either crooked edges on the finder patterns or uneven spacings within the alternating timing patterns running between the finder patterns and aligned with the inner boundaries of these.

The finder patterns and the adjacent quiet zone areas should always be solidly dark and light. Failures in the print mechanism which may produce defects in the form of light or dark streaks through the symbol should be visibly evident where they traverse the finder pattern or the quiet zone. Such systematic failures in the print process should be corrected.

M.5 Assessing print growth

A linear bar code verifier capable of outputting direct measurements of bar and space patterns may be used for the assessment of print gain or loss in both horizontal and vertical axes, by measuring along two scan paths at right angles passing through a finder pattern and crossing the center 3 x 3 block of modules. Analysis of the output should reveal an apparent bar/space/bar/space/bar pattern; the print gain (or loss) can be assessed by comparing the five measured element widths with the ideal 1 : 1 : 3 : 1 : 1 ratio of the widths.

Annex N
(informative)

Characteristics of Model 1 symbols

N.1 Model 1 QR Code symbols

Model 1 of QR Code, as defined in AIM ITS 97-001, is the form of the symbology originally used for a number of early or closed systems applications but is not recommended for use in new or open systems applications, and is unsuitable where data volumes are likely to be high. In most respects it follows the same specification as QR Code but differs in a number of significant aspects which are summarised in this Annex.

N.1.1 Model 1 overall characteristics

a) Symbol size (not including quiet zone):

   21 × 21 modules to 73 × 73 modules (Versions 1 to 14, increasing in steps of 4 modules per side).

b) Maximum data capacity (for maximum symbol size with lowest level of error correction, Version 14-L):

   — numeric data: 1 167 characters

   — alphanumeric data: 707 characters

   — Byte data: 486 characters

   — Kanji data: 299 characters

c) Symbol structure and features compared with QR Code:

   — alignment patterns: Model 1 symbols have no alignment patterns

   — extension patterns: Model 1 symbols have extension patterns on the right-hand and lower sides

   — version information : Model 1 symbols contain no version information

   — symbol character placement : in consequence of the above, symbol character placement follows different rules.

   — Model 1 symbols do not support the ECI protocol

   — Model 1 symbols do not support mirror imaging

   — Model 1 symbols do not support reflectance reversal

d) Error correction: the error detection and correction codewords are calculated identically with QR Code, but the number and size of error correction blocks for any Version differ. Data and error correction codeword blocks are not subject to interleaving.

Figure N.1 below illustrates the structure of a Version 7 Model 1 QR Code symbol.

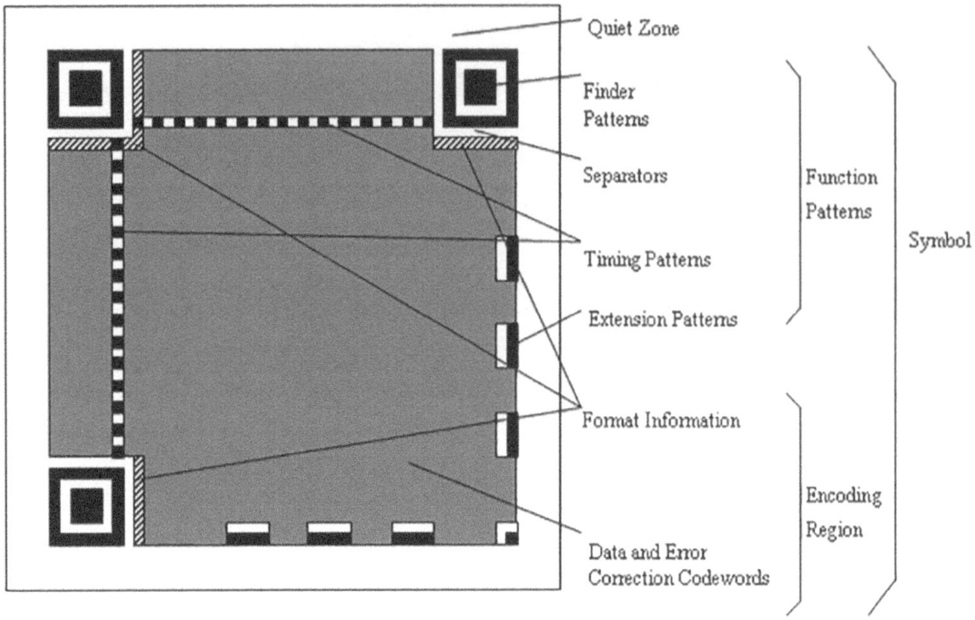

Figure N.1 — Structure of a QR Code Model 1 symbol

N.1.2 Symbol versions and sizes

There are only fourteen sizes of Model 1 symbol, from Version 1 to Version 14, the sizes of which are identical with those of Model 2 symbols with the same Version numbers, as defined in 6.3.2. Version 1 symbols therefore measure 21 × 21 modules, and Version 14 symbols 73 × 73 modules. Table N.1 shows the data capacity of all Model 1 symbols at the different error correction levels.

Table N.1 — Data capacity of all versions of Model 1 QR Code

Version	No. of Modules/side (A)	Function Patterns Modules (B)	Format Information Modules (C)	Data Modules except (C) (D=A^2-B-C)	Data Capacity [codewords]* (E)
1	21	206	31	204	26
2	25	230	31	364	46
3	29	238	31	572	72
4	33	262	31	796	100
5	37	270	31	1 068	134
6	41	294	31	1 356	170
7	45	302	31	1 692	212
8	49	326	31	2 044	256
9	53	334	31	2 444	306
10	57	358	31	2 860	358
11	61	366	31	3 324	416
12	65	390	31	3 804	476

Table N.1 *(continued)*

Version	No. of Modules/side (A)	Function Patterns Modules (B)	Format Information Modules (C)	Data Modules except (C) (D=A^2-B-C)	Data Capacity [codewords]* (E)
13	69	398	31	4 332	542
14	73	422	31	4 876	610

NOTE The first codeword is 4 bits in length. All subsequent codewords are 8 bits in length. The first, 4-bit, data codeword is prefixed with 0000 to make its length 8 bits for generating the error correction codewords.

N.2 Detailed specifications

For complete information regarding the printing and reading of Model 1, see AIM ITS 97-001.

Bibliography

[1] ISO/IEC 646, *Information technology — ISO 7-bit coded character set for information interchange*

[2] ISO/IEC 8859-2:1999, *Information technology — 8-bit single-byte coded graphic character sets — Part 2: Latin alphabet No. 2*

[3] ISO/IEC 8859-3:1999, *Information technology — 8-bit single-byte coded graphic character sets — Part 3: Latin alphabet No. 3*

[4] ISO/IEC 8859-4:1998, *Information technology — 8-bit single-byte coded graphic character sets — Part 4: Latin alphabet No. 4*

[5] ISO/IEC 8859-5:1999, *Information technology — 8-bit single-byte coded graphic character sets — Part 5: Latin/Cyrillic alphabet*

[6] ISO/IEC 8859-6:1999, *Information technology — 8-bit single-byte coded graphic character sets — Part 6: Latin/Arabic alphabet*

[7] ISO/IEC 8859-7:2003, *Information technology — 8-bit single-byte coded graphic character sets — Part 7: Latin/Greek alphabet*

[8] ISO/IEC 8859-8:1999, *Information technology — 8-bit single-byte coded graphic character sets — Part 8: Latin/Hebrew alphabet*

[9] ISO/IEC 8859-9:1999, *Information technology — 8-bit single-byte coded graphic character sets — Part 9: Latin alphabet No. 5*

[10] ISO/IEC 8859-10:1998, *Information technology — 8-bit single-byte coded graphic character sets — Part 10: Latin alphabet No. 6*

[11] ISO/IEC 8859-11:2001, *Information technology — 8-bit single-byte coded graphic character sets — Part 11: Latin/Thai alphabet*

[12] ISO/IEC 8859-13:1998, *Information technology — 8-bit single-byte coded graphic character sets — Part 13: Latin alphabet No. 7*

[13] ISO/IEC 8859-14:1998, *Information technology — 8-bit single-byte coded graphic character sets — Part 14: Latin alphabet No. 8 (Celtic)*

[14] ISO/IEC 8859-15:1999, *Information technology — 8-bit single-byte coded graphic character sets — Part 15: Latin alphabet No. 9*

[15] ISO/IEC 8859-16:2001, *Information technology — 8-bit single-byte coded graphic character sets — Part 16: Latin alphabet No. 10*

[16] ISO/IEC 15416, *Information technology — Automatic identification and data capture techniques — Bar code print quality test specification — Linear symbols*

[17] ISO/IEC 15417, *Information technology — Automatic identification and data capture techniques — Code 128 bar code symbology specification*

[18] ISO/IEC 15424, *Information technology — Automatic identification and data capture techniques — Data Carrier Identifiers (including Symbology Identifiers)*

[19] ISO/IEC 15434, *Information technology — Automatic identification and data capture techniques — Syntax for high-capacity ADC media*

[20] ISO/IEC/TR 29158, *Information technology — Automatic identification and data capture techniques — Direct Part Mark (DPM) Quality Guideline*

[21] *AIM ITS 97-001 International Symbology Specification - QR Code*, AIM Inc.

[22] AIM International Technical Specification, Extended Channel Interpretations:—Part 1, Identification Schemes and Protocols

[23] AIM International Technical Specification, Extended Channel Interpretations:—*Part 2, Registration Procedure for Coded Character Sets and Other Data Formats*

[24] AIM International Technical Specification, Extended Channel Interpretations:—Character Set Register

[25] *GS1 General Specifications*, GS1

[26] JIS X 0208:2012, *7-bit and 8-bit double byte coded KANJI sets for information interchange*

ICS 01.080.50; 35.040
Price based on 117 pages

Made in United States
Cleveland, OH
03 November 2026

47601688R00069